Charles Seale-Hayne Library
University of Plymouth
(01752) 588 588
LibraryandITenquiries@plymouth.ac.uk

MULTIMEDIA TELECOMMUNICATIONS

JOIN US ON THE INTERNET VIA WWW, GOPHER, FTP OR EMAIL:

WWW: http://www.thomson.com
GOPHER: gopher.thomson.com
FTP: ftp.thomson.com
EMAIL: findit@kiosk.thomson.com

A service of I(T)P®

BT Telecommunications Series

The BT Telecommunications Series covers the broad spectrum of telecommunications technology. Volumes are the result of research and development carried out, or funded by, BT, and represent the latest advances in the field.

The series includes volumes on underlying technologies as well as telecommunications. These books will be essential reading for those in research and development in telecommunications, in electronics and in computer science.

1. *Neural Networks for Vision, Speech and Natural Language*
   Edited by R Linggard, D J Myers and C Nightingale

2. *Audiovisual Telecommunications*
   Edited by N D Kenyon and C Nightingale

3. *Digital Signal Processing in Telecommunications*
   Edited by F A Westall and S F A Ip

4. *Telecommunications Local Networks*
   Edited by W K Ritchie and J R Stern

5. *Optical Network Technology*
   Edited by D W Smith

6. *Object Oriented Techniques in Telecommunications*
   Edited by E L Cusack and E S Cordingley

7. *Modelling Future Telecommunications Systems*
   Edited by P Cochrane and D J T Heatley

8. *Computer Aided Decision Support in Telecommunications*
   Edited by P G Flavin and K A E Totton

9. *Multimedia Telecommunications*
   Edited by W S Whyte

MULTIMEDIA TELECOMMUNICATIONS

Bill Whyte
School of Computer Studies
University of Leeds
UK

CHAPMAN & HALL

London · Weinheim · New York · Tokyo · Melbourne · Madras

Published by Chapman & Hall, 2–6 Boundary Row, London SE1 8HN, UK

Chapman & Hall, 2–6 Boundary Row, London SE1 8HN, UK

Chapman & Hall GmbH, Pappelallee 3, 69469 Weinheim, Germany

Chapman & Hall USA, 115 Fifth Avenue, New York, NY 10003, USA

Chapman & Hall Japan, ITP-Japan, Kyowa Building, 3F, 2-2-1 Hirakawacho, Chiyoda-ku, Tokyo 102, Japan

Chapman & Hall Australia, 102 Dodds Street, South Melbourne, Victoria 3205, Australia

Chapman & Hall India, R. Seshadri, 32 Second Main Road, CIT East, Madras 600 035, India

First edition 1997

© 1997 British Telecommunications plc

Printed in Great Britain

ISBN 0 412 78600 1

A catalogue record for this book is available from the British Library

∞ Printed on permanent acid-free text paper, manufactured in accordance with ANSI/NISO Z39.48-1992 and ANSI/NISO Z39.48-1984 (Permanence of Paper).

To Marian, William and Alasdair

April 1997

Contents

Contributors

R A Bissell	Interactive Multimedia, BT Laboratories
W Bunn	LAN Applications, BT Laboratories
P M Clarke	Advanced Media, BT Laboratories
J W Cook	Copper Access Systems, BT Laboratories
W Dobbie	Broadcasting and Interactive TV, BT Laboratories
A Eales	On-line Platforms, BT Laboratories
M D Eyles	Multimedia Evolution, BT Laboratories
M Fauth	Shared Information Environments, BT Laboratories
K T Foster	Transmission Technologies, BT Laboratories
M J Gray	Formerly Teleworking, BT Laboratories
E W Jones	Multimedia Services, BT Laboratories
G W Kerr	Interactive Multimedia Services, BT Laboratories
T Midwinter	Multimedia Development, BT Laboratories
J Morphett	Shared Information Environments, BT Laboratories
P A Rea	Shared Information Environments, BT Laboratories
A S Rogers	Formerly Virtual Reality Environments, BT Laboratories
D A D Rose	Advanced Media, BT Laboratories
G L Smith	Multimedia Services, BT Laboratories

D A Tilson	Formerly Customer Products, BT Laboratories
J A Totty	Multimedia Evolution, BT Laboratories
G R Walker	Shared Information Environments, BT Laboratories
W S Whyte	School of Computer Studies, University of Leeds (formerly BT Laboratories)
C D Woolf	Customer Products, BT Laboratories
G Young	Copper, Radio and Satellite Systems, BT Laboratories

Preface

When I was asked to be editor of the BT Technology Journal themed issue of October 1995 on 'Multimedia Communications', and then subsequently of this book, I was aware the assignment comprised two formidable tasks — attempting to cover in one volume, an immensely vast and diverse field of inquiry and, moreover, one that was changing daily.

My approach to the former, then and now, was to omit CD-ROM and other non-communications related approaches, images and speech processing and multiservice communications networks, which are all well covered elsewhere. Instead, I attempted to focus on material with a specific communications orientation, trying to produce a selection that gave some idea of the diversity of activity going on under the title, all the time trying to pull it together using the common thread of the 'human dimension', which I maintain is the principal shaper of the various different forms of multimedia communication.

As far as the pace of change is concerned, I have called upon the good nature of the contributors, in the hope that they would not let me down. In this, at least, my judgment and their generosity have not faltered, as most authors have revised, in some cases extensively, the material that first appeared in the BT Technology Journal (October 1995) issue. To them, many thanks.

I have also, in this book, given greater recognition to the importance of the Internet, though many chapters remain concerned with the high-end of 'seductive' multimedia (see my introductory chapter for a fuller description of this). The image and sound attributes of video-on-demand technology or digital broadcast are areas where the quality they deliver still exceeds to a greater degree anything that can be offered via the Internet. However, the latter has a vibrancy that should not be ignored and I have included a chapter on Internet capability and a discussion on its strengths and weaknesses. (I am particularly grateful to the Editor of the British Telecommunications Engineering Journal, for permission to publish this.)

Of course, the Internet has also had a major influence on the dissemination of knowledge — its somewhat anarchic principles greatly accelerate the publication of new results and demonstrations, provided you know where to look. To this end, the url addresses of some Web sites are included in the References to the various chapters. They are like most Internet material — quirky and sometimes unavailable. This is part of the excitement of exploring the future, and therefore very appropriate to multimedia communications, at the very root of whose success is its appeal to human emotion, interest and, above all, 'fun'.

Bill Whyte
School of Computer Studies, University of Leeds
e-mail:billw@scs.leeds.ac.uk

1

THE MANY DIMENSIONS OF MULTIMEDIA COMMUNICATIONS

W S Whyte

1.1   INTRODUCTION

The many facets of multimedia communications are dictated by the compromises that technology has to make in attending to a range of varying human needs — emotional satisfaction or judgment support — that must be met. Given that no universal solution addresses all of the requirements, it is natural that solutions for entertainment, for example, that use asymmetrical broadband transmission, differ from the symmetrical, switched, more modest bandwidth solutions for business applications such as distributed co-operative working. This introduction aims to position the chapters that follow within this human-centred framework.

Looking first at content-driven services such as movies-on-demand and home shopping, the processes required to fit the content to the available delivery platforms are examined in ways that encourage reuse and minimize repeat costs. Then a number of delivery methods are considered — digital TV, video-on-demand servers and set-top boxes, and the transmission technology for broadband delivery to the home.

Also covered are public multimedia terminals, which meet the triple requirements to provide 'seduction, information and transaction', by mixing locally stored, high-quality images with information updates and purchasing transactions carried over basic-rate ISDN.

Moving to co-operative uses of multimedia, the requirements of 'decision environments' are considered, giving examples of teleworking and desktop multimedia. The issue of standardization is clearly important, as is the integration with existing non-multimedia office automation infrastructures, and between computing and telecommunications.

Virtual environments, built round the logical realities of co-operative activities, rather than the accidents of geography, can assist in problem solving and conceptualization, and some examples are discussed.

Although it does not yet provide high-quality moving image or audio, the Internet is a vibrant and extensive source of content and some of the recent developments in the area of graphics are examined.

Communications techniques are the key to opening up many markets for multimedia, but there are a number of options, each with its own advantages and problems. A variety of ways of delivering entertainment services are looked at, particularly those where the requirement is predominantly for high bandwidth in one direction only, and also considered is the particular problem that arises with variable transmission delay on computer networks, especially in the case of the Internet.

1.2   WHAT IS MULTIMEDIA COMMUNICATION?

The common complaint that begins most discussions on the subject of multimedia is that the biggest problem lies in the definition of the term, or at least in trying to reach some form of consensus about its scope. Some include television, some exclude videotelephony, some would do the reverse. Others have different definitions, which are all rational but frequently mutually exclusive.

Maybe this is because multimedia, and, even more, multimedia communications, is an entity of many dimensions. For example, it can be described in terms of its technology platforms, their bit rates and connectivity, its vendor markets (communications, computing, consumer electronics), its service attributes (interactive, store-and-forward, volatile, static, etc) and its applications. Later chapters in this book provide a wide, and therefore sparse, set of points in this multidimension. This introduction will try to plot some of the connecting threads that span the multimedia dimensions, placing the other chapters within this connectivity.

Of all these dimensions, we would claim that the principal axis of multimedia communication, the single dimension that best explains its complexity, is its relationship to the human dimension, i.e. how we sense, reason and feel.

1.3   DEFINITION

Multimedia communications aspires to be a 'worm-hole in space', putting all our perceptions, our understanding and our emotions in direct contact with distant environments, and they with us, but, even in best practice, it is a window of distorting glass that blocks out most of touch and all of smell, and through which light and sound travel slowly and usually preferentially in a particular direction.

Compromises are therefore required, to meet the specific purpose and, with all this in mind, the following working definition is offered:

'Multimedia communications, through a range of technologies, including communication over a distance, aspires to provide a rich and immediate environment of image, graphics, sound, text and interaction, which assists in decision-making or emotional involvement.'

This definition recognizes the purpose of multimedia communications — it is a support tool to human activity, an activity which can come in different forms, in particular, emotionally and intellectually.

It suggests that it is unlikely that these needs will be met by a single technology solution.

Also, it adopts the emerging consensus that there are a limited number of media types (text, etc) that are practically available. It is silent regarding whether the presence of any specific media type is a prerequisite for the term 'multimedia' to apply, although it is noted that there is a school of thought that considers image, preferably moving image, to be such a requirement.

It introduces the concept of 'interaction', i.e. the ability of the user to have some significant control over the virtual environment to which she or he is exposed. This requirement seems to have crept into common use to distinguish 'true' multimedia from simpler systems such as broadcast television or video cassette. In practice it is a useful distinction, but it must not be used to obscure the fact that broadcast TV provides a benchmark for professionalism of content and quality of transmitted image and sound against which, in the entertainment arena at least, all other multimedia offerings will be judged.

Finally, the definition recognizes that distance between the user and the real world poses specific problems which can be alleviated by communications. It is true that multimedia can be provided on a stand-alone basis — consider two simple examples: the use of CD-ROM as a convenient and inexpensive way of accessing still or moving images, and the creation of virtual worlds on stand-alone desktop computers — but in many applications there is a need to update the data quickly and regularly, for people at a distance to work together on the same set of images, or for entertainment services to be distributed to millions of homes. This is where the communications comes in and where solutions which are currently limited in application because of their stand-alone nature may be released into a wider market.

1.4   SOME APPLICATIONS

At this point it may be useful to consider briefly some concrete applications of multimedia communications. The list that follows does not attempt to be exhaustive:

- training and education — the ability to provide access to remote databases and to interact, 'face-to-face', with a tutor located elsewhere;

- distributed work groups — creating a logically unified project team, with common access to multimedia data, e.g. engineering designs and videoconferencing;

- remote experts — providing images from inaccessible locations, e.g. oil rigs, to headquarters' experts who can, in return, communicate advice and assembly diagrams;

- information/sales terminals — located in shops, departure lounges, public places, providing information on products or services, perhaps with videotelephony access to a salesperson or customer service;

- interactive television — entertainment, shopping, education and information services based around domestic television and consumer/product interfaces, e.g. hand-held remote controller;

- computer-based information services — as the name implies, built around personal computer access to information services, e.g. Internet.

For a more comprehensive description and extensive bibliography, see Williams and Blair [1].

1.5   THE HUMAN DIMENSION

Multimedia communications has been defined as being driven by the requirements of its users. Given that the present state of technology will require trade-offs, it is important to identify and segment the principal elements of these requirements, in human terms.

1.5.1   Judgment and emotion

Of course, almost all human activities are a mixture of the intellectual and the emotional, but insofar as they can be separated out in any one activity type, they lead to significantly different requirements. This may be illustrated by a specific example.

Consider home shopping by means of a catalogue — the catalogue can be analysed into three components:

- the illustrations, photographs of the goods, set in attractive surroundings perhaps modelled by flawless models, the prime purpose being to entice you to buy the goods — let us call this the 'seduction' component;

- beside the illustrations are descriptions of the goods, their prices, colours, sizes, etc, and sometimes a diagram giving details of the construction or usage of the product — the prime purpose here is to provide an 'information' component;

- finally, somewhere in the catalogue will be an order form with a postal address or a telephone number, to allow you to place your order — this represents the 'transaction' component.

These components are relevant to paper-based, catalogue shopping and to its more modern equivalent, 'electronic shopping'. Using the terminology of the latter, these three 'requirement' components are mapped, in Fig. 1.1, on to the 'solution' components that are required in order to deliver them:

- multimedia;

- intelligent databases;

- communications.

requirement / solution	seduction	information	transaction
multimedia	exciting, high quality moving images and sound, animated graphics face-to-face videophony	graphics, stills, spoken instructions	
databases		hold content and results of interactions but do not directly interact with people	
communications	access to up-to-the-minute/live feeds, remote assistants, counsellors	access to remote information, updates, remote experts	order placing, payment taking, delivery of electronic goods, negotiations with remote sales points

It should be noticed that the multimedia component has a major contribution to make to the seduction element and also a significant one to the information part, reinforcing the implication of our definition that multimedia affects emotion, that is, 'seduction', and also assists decision-making, the 'information' part. The multimedia component in the seduction element aspires to be exciting, vivid, movie quality, coloured, audibly rich; the multimedia component in the information domain should be clear, informative, perhaps an animation, but not necessarily moving video. There is almost certainly an abundance of text and/or line diagrams. (This is not really a new discovery — there is a well-known adage in publishing regarding the use of colours: 'Two to tell and four to sell'.) Here the current strengths and weaknesses of the Internet are revealed — its information content is immense, but its presentation of them, decidedly creaky.

In general, the required richness of media for decision-making is less than that for emotional involvement.

It must not be assumed that moving picture quality is the single dimension of overall emotional stimulation. People often need fast, two-way interaction to achieve emotional satisfaction, e.g. in achieving empathy. The one-way transmission of images, from machine to human, does not fully meet this need, nor does a message-based, person-to-person video link. Sometimes conversational voice or videotelephony is required.

Furthermore, an informal body of evidence is emerging, in conversations with advertising and media bodies, that the quality of audio, in terms of its content and its recording and reproduction, is a major element in providing emotional satisfaction.

The contribution of audio to the multimedia environment is an under-researched, or at least under reported, area of study.

1.5.2   The asymmetry of content

Even if media quality, in terms of clarity of image and sound were a necessary requirement for a satisfactory emotional experience, they would not, on their own, be sufficient; to paraphrase slightly a technical director of Sega, the computer games company: 'The games platform is only an enabler; apart from price, three further things are required: content, content, content.' [2]

The creation of high-quality content, either for games, TV or any other entertainment service intended to capture our emotions, requires rare talent and considerable expense. This applies both to the performers in multimedia entertainment and to the technical and design people involved in putting together the structure that surrounds the performance, whether it be a movie, home shopping, banking or other service. Chapter 2 describes a typical service creation environment for the generation of such offerings.

Specifically, Chapter 2 distinguishes 'service design', i.e. the generic infrastructure, from 'application creation', i.e. the design and creation process which is undergone anew for each 'shop', 'bank', 'cinema-on-demand' offering. Service design encompasses the commercial issues of the scope and scale of the service, including how it is to be branded and what ordering, payment and other transactions are permitted, technical issues such as platform type and media capability, and user interface issues such as generic rules for navigation and screen layouts.

Completion of the service design results in the creation of a number of documents which are used in the application creation stage. This addresses the construction of specific offerings from individual retailers. Typically, it makes extensive use of storyboards and workshops involving creative designers and people who understand the capabilities and limitations of the platform.

This process can be described as: 'Trying to produce a fun offering, on time, on budget'. More graphically, it can be described as: 'Getting the Bolshoi Ballet (the creative artists) and the Red Army (the database specialists and systems programmers) to work together under control of the KGB (the management)'!

It follows directly from the difficulty and expense of creating absorbing multimedia material that there will be a decided tendency for there to be more receivers of quality content than there are transmitters.

As a corollary of this, it follows that content-based multimedia services will be highly 'asymmetric' — there will be a requirement for large data flows in one direction and not in the other. Also, it becomes cost effective to place the burden of cost on the relatively few production equipments and aim for low-cost receivers, as will be seen when video on demand is discussed. Related to this asymmetry, it also follows that the content will be created and stored off-line for access by the user at the most suitable time.

These statements naturally lead to a store-and-retrieve architecture, rather than a fully bi-directional, low end-to-end delay service. Thus, there will be a tendency to support standards such as MPEG [3, 4] media encoding, with its high complexity of coding, low complexity decoding, and its relatively long buffering requirements, rather than the H.261 standard for videophony.

In order to capitalize on the reuse of expensive content, its creators will attempt to sell their creations across a number of platforms — CD, TV, etc. Two consequences follow directly from this.

- Firstly, a standards issue — developments in the computing, consumer electronics and entertainment industries can be expected in order to seek standardization at the video and audio encoding level, however reluctant they are to lose differentiation [5]. The Betamax/VHS wars haunt the industry and it is unlikely that domestic consumers will tolerate such a thing again. It has been interesting to observe the agonizing undergone by the players in the consumer electronics market over compact disk standards [6].

- Secondly, there will be a need for platform-independent 'service creation tools', to make it easy for the application industry to author new applications across a variety of platforms. A good example of this has been the BT Laboratories development of in-flight information, communication and entertainment services, that operate on a number of different on-board operating systems and hardware, provided by each airline. Figure 1.2 shows the layered nature of an easily portable creation environment.

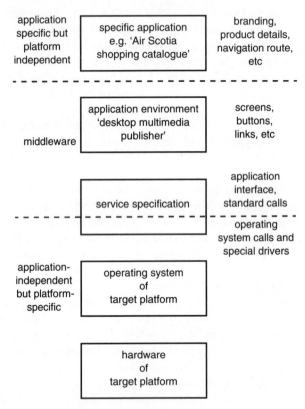

Fig 1.2   The layered nature of a portable creation environment.

At the top of this environment is the specific application, a shopping catalogue for an imaginary airline. This will have been designed through the story-boarding process described in Chapter 2. This will be mainly the work of the content and service provider (or their agents) and embody the brand-image, the way they want to show product details, and so on.

This application, as with all applications developed in the environment, will be constrained within certain rules, regarding number of media windows, position and size of buttons, page links, etc, through an 'application design environment', essentially a multimedia publishing package. This environment will, in general, be independent of the platform on which the application will run (although specific extensions may be permitted in some circumstances).

This platform independence has been achieved by the identification of a sufficient set of standard procedures that cover the range of activities required to produce the repertoire of applications — procedures such as, 'fetch page n', 'link image to window y', etc.

Armed with this application environment and a storyboard, the application developer can create the application without knowing or caring about any details of the platform.

The developer of the application environment, however, in addition to creating the standard, platform-independent application procedures, also has to understand the details of the operating system and hardware of the target platform. Thus the developer must produce a piece of middleware that converts between the generic procedure calls of the application environment (not the application itself) and the realities of the drivers and utilities of the operating system, sometimes even writing directly to the hardware.

Developing this application environment and middleware costs time and money, up-front before any actual application is delivered, and requires foresight and understanding from the customer. However, it is undoubtedly the correct course of action.

1.5.3   Multimedia structures

Another aspect of standardization is covered in Chapter 3, which considers the handling of multimedia content from a variety of source types, generated and presented on a variety of hardware platforms.

It is argued in Chapter 3 that content provision will be a major source of both revenue and costs; this will give rise to a strong desire for reuse and the solution will be to adopt a standard, intermediating format for the transfer of multimedia content. One such format has been addressed by the Multimedia and Hypermedia Information Coding Expert Group (MHEG) in ISO. Data structures held within platforms and applications and coded in private or proprietary format can be recoded and decoded to and from a standard MHEG format, for transferring to the presentation device.

Clearly, the MHEG coding scheme must be flexible enough to support most, if not all, foreseeable multimedia presentations. The coding scheme should support the referencing and/or the encapsulation of objects encoded through existing format such as MPEG video, G.711 audio, etc. It also should allow the composing of complex objects and expressing relationships, such as spatial and temporal position (e.g. subtitles on a moving video), and so on. Adopting this approach, it is argued, leads to a better understanding of the distinction between content and application, and removes the reliance on a hardware/software 'stovepipe' solution.

1.5.4   Interactive TV

So far, this section has identified a number of system attributes to meet the needs of emotionally satisfying, high-quality content, particularly with a domestic and entertainment slant:

- good media quality;

- cheap receivers;

- asymmetrical distribution;

- store and access;

- standardized construction.

Recently, the label 'interactive TV' has become applied to a range of products that go some way to meeting this list of requirements. They include broadcast satellite, terrestrial and cable TV with some form of backward channel, perhaps PSTN or packet data, and also the emerging technology of video-on-demand.

1.5.5   Broadcast television

In Chapter 4, Dobbie claims that: 'Broadcast television is perhaps the most widely accepted form of multimedia communication today'. The major change from analogue to digital is imminent and, along with this change, comes the opportunity to provide a range of levels of personalization of service, including backward channels that allow user interaction.

After reviewing current standards, he describes how digital transmission has evolved — the broadcast coding standard has stabilized on MPEG-2, incorporated within a set-top box with separate inputs for terrestrial, satellite and cable delivery. There will also be 2-way connectivity to PSTN or ISDN which will provide the backward channel as well as a means of delivering information services, e.g. Internet, or the downloading of software for games, etc.

Eventually the functionality of the set-top box is likely to be incorporated within the TV set itself, given suitable markets and agreed standards.

Clearly, the two issues are interrelated and the roles of the two key drivers of standards — the European Digital Video Broadcasting Project (DVB), and the Digital Audio-Visual Council (DAVIC) — are discussed, as are a number of specific issues and agreements.

Membership of these two bodies demonstrates the role of multimedia communications enforcing convergence upon the computing, consumer electronics and telecommunications industries.

1.5.6   Video on demand

The basic architecture of video on demand is shown in Fig. 1.3. Video material — films, TV programmes, home-shopping material, etc — is stored as compressed digitized material on a video server. The servers are accessed by set-top boxes (on top of domestic TV sets), on demand of the user, who typically controls the set-top box with a standard TV remote control unit. Requests from the set-top box travel over a relatively slow-speed backward channel to the server which can reply with a continuous stream of data at megabits per second rates and unique to the user — hence the 'on-demand' name.

Each of the components of the service sets some significant technology challenges.

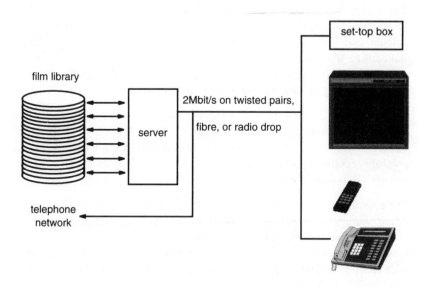

Fig 1.3   Video-on-demand basic architecture.

- Video servers

  In Chapter 5, Kerr describes the issues surrounding the design of the video server. Although he also covers some aspects of 'low bit rate' servers operating at fractional Mbit/s rates, the main thrust of the chapter is on equipment capable of continuously delivering bit rates in excess of 2 Mbit/s per user. Given the large-scale demands for data storage and access — a video library of 2000 hours of material requires 2 Tbytes capacity and 2 Gbit/s access speeds to cover over two thousand simultaneous accesses — it is not surprising that network planning and trade-off between server location and size are major issues.

  Service architecture is also heavily dependent on application; passive viewing places fewer demands than do interactive services such as home shopping.

  The chapter reviews the different designs adopted by vendors. Some use special-purpose hardware, some employ massively parallel processors, while another school of thought believes that multiple small-scale conventional processors offer the low-cost, expandable solution.

- Set-top boxes

  The set-top box architecture and evolution are described in Chapter 6. The set-top box is a consumer item whose success is critically dependent on the amount and quality of content available. Consequently, it is essential to standardize the format of the content, as far as possible, across different manufacturers of the same application platform, and also across different applications, e.g. set-top boxes for video-on-demand and satellite.

  The set-top box used in the BT video-on-demand trial is an Apple Macintosh LC-series computer. Chapter 6 describes the 'CPU-centric' design architecture, both software and hardware. It then goes on to predict possible evolution paths for the set-top box, including its ultimate absorption into the TV set.

  The authors believe that the overall design principles will be enduring, with minor variations depending on the split between hardware and software implementations of functional blocks. However, the market will split into at least two classes of box — a basic unit for simple interactive TV capability, with minimal downloading of complex applications, and a high performance unit, based on computer workstation designs, capable of considerable expansion and very high resolution graphics.

An exciting development in the set-top box debate, is the question of the partitioning of executable software and operating system between the server and the set-top box. Networked software such as Java [7] and Telescript [8] work on the principle of sending complex files from network servers to office or home terminals such as personal computers or set-top boxes. These files contain not just data but also executable code, which can run locally on the terminal, after it has been compiled or, more usually, interpreted, by the terminal. One implication of this, it is claimed, is the possibility of providing 'stripped-down' personal computers that require much lighter operating systems and associated memory and processing power, and thus reduced cost. There are industry debates about whether this is realistic, but these are based as much on market manoeuvring as technical fact. However, the emergence of network code will undoubtedly enable the more rapid installation and maintenance of changes to set-top box and personal computer software. This could include service control software, such as billing and access approval to subscription services, games software, special effects and new image or audio coding schemes.

- Delivering video-on-demand to the home

In terms of data quantity and data rate, video-on-demand clearly sets some exacting requirements.

There are a number of transmission techniques under development for delivering the data to the customer's premises. 2 Mbit/s radio is a possibility, as is fibre, but perhaps the most interesting is the so-called 'asymmetric digital subscriber loop' (ADSL) method of delivery over the existing copper pair. Chapter 7 explains how data rates of several megabits per second over multiple kilometre lengths of unmodified cable have been successfully demonstrated in the field. The 'trick' is simple: by restricting the transmission of high-speed data to one direction, from exchange to customer, there is no problem with near-end cross-talk, which is usually the factor that determines the maximum range in the case of bi-directional transmission.

Even so, it is not possible to achieve acceptable error rates without making use of complex coding and equalizing techniques and a number of these exist without a clear winner identified at this time. This is a very active area of investigation and hardware designers can take heart that, beside the digital and signal processing developments, are some very demanding 'real' filter design problems in separating the high-frequency VoD signals from the low frequency telephony ones.

ADSL is now a proven technology, with rate-adaptive chipsets becoming available, which, in some cases, can be configured to operate from a few tens/ hundreds of kilobit/s downstream to 8 Mbit/s downstream and up to 1 Mbit/s upstream, depending on the length of the copper loop.

Originally, ADSL developments concentrated on exchange-to-subscriber systems, but more recently there has also been an interest in shorter range, wider bandwidth solutions that operate between distribution points, such as kerb-side cabinets, and the customer premises. (In this case, the exchange-to-cabinet connection is by means of optical fibre).

1.5.7   Emotion, decision, commitment — public terminals

The simplest application for interactive TV is probably the selection of movies or television recordings, but it is also easy to see how it could be extended to provide more complex interactive services. One such example is the home shopping scenario, with the triple need to engage, inform and transact with the customer. Interactive television is an obvious vehicle for achieving this, but currently there remain barriers in the way — broadcast and many cable services lack sufficient ability to be personalized to individual requirements; video on demand is still experimental and is currently range limited.

What is required is the media quality of interactive TV with the ubiquity of PSTN.

In the discussion of public terminals in Chapter 8, some compromises and synergies in this direction are explored.

The terminals are located in places visited by subsets of the public — customers in a shop, travellers on an aeroplane, in an airport concourse, visitors to a museum, etc. Sometimes they sell goods or try to provide an absorbing experience, sometimes they are more concerned with providing information such as street plans or community services, or providing access to remote expertise.

Terminals can be 'point of information', providing information in predefined order, 'browsers', giving freedom to roam through the database, 'option selectors', where assemblies of goods and services can be created and order tickets produced, and 'full function' kiosks which provide a telecommunications channel to a remote customer-handling point, together with the ability to handle financial transactions and presentation of receipts.

It is contended in Chapter 8 that the natural evolution path is in the direction of 'full functionality'. This is the only way that high-quality images can be combined with a mechanism for committing the customer to a transaction, for example, a purchase.

Traditionally, terminals held their high-quality images locally on mass storage devices such as hard disk or CD-ROM, but this presents some problems, notably the difficulty of issue control if a large number of terminals are involved, and the need to update the database, for example, with product price and availa-bility. This is one area where a communications link can help, even though broadband networks, such as video-on-demand, are not yet widely available. It can also offer the possibility of a 'pop-up' assistant, whose eye-to-eye involve-ment with the potential customer may facilitate a sale.

A number of options are explored in Chapter 8 for combining the switched, flexible and controlled network connection with local processing and storage. The chapter also expands on the requirements of kiosks to handle transaction cards, the need for network management and maintenance control, and how they can be integrated into a complete fulfilment operation.

1.5.8   Utility versus emotion

When the requirements and the solutions so far are compared, a clear pattern emerges — highly absorbing, quality content, with vivid images and sound, can only be delivered at the expense of distance, bi-directional bandwidth, immediacy of generation and flexibility of origin. Figure 1.4 shows this in a rather idealized form.

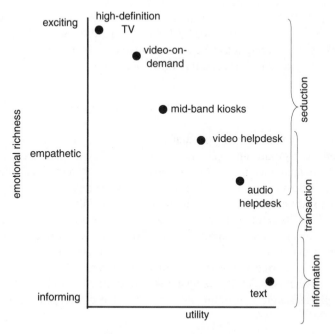

Fig. 1.4   Utility versus emotion.

The vertical axis represents the emotional impact produced by the media. As has been shown, this is highly correlated with media richness and thus bandwidth. (This can be one-way richness, for entertainment services, but both-way bandwidth and latency must be taken into account for empathy-based services, such as conversations.)

The horizontal axis attempts to depict a rather abstract and ill-defined quantity, 'utility', which is to be interpreted as a complicated function of price, usability, ubiquity of access and connectivity between multiple receivers and multiple sources. It is not an exact quantity, but it has proved of practical value in discussions, with most people being consistent in their placing of products and services along the axis, at least when discussing similar applications. It appears in a slightly different form in Chapter 8.

Superimposed on Fig.1.4 are the three zones of seduction, transaction, information. For example, a telephone helpdesk transaction need not be exciting, but there must be a reasonable degree of empathy between the agent and the customer. This would be enhanced by the presence of videotelephony, but, of course, current pricing and availability would then reduce the utility.

In general, the goal of multimedia communications is the top, right-hand corner, but different requirements are necessarily positioned at different points from it. As previously described, entertainment services, benchmarked against broadcast TV, have had to sacrifice utility in terms of range (video-on-demand), connectivity (CD-ROM) and interactivity (satellite and video-tape) for media richness; kiosks have compromised on emotion and utility, but can achieve a top-right shift by introducing a transmission component, such as a basic-rate ISDN connection, for video updates and real-time interaction.

1.5.9   Decision environments

Multimedia communications potentially allows us to integrate a number of distributed work-groups into a single, logical office, but any attempt to introduce multimedia communications into the business environment must accept that businesses demand a very high utility from all their processes. Specifically, anyone who has been involved in the introduction of office automation knows that major benefits can only be achieved when a practically ubiquitous service is provided. It is much more important to provide basic e-mail to everyone in the organization than making available compound, modifiable documents only to a select few. In fact, Metcalfe (one of the pioneers of Ethernet) equates the 'value' of a network (that is, its utility) to the number of users squared [9].

This places severe constraints on the realization of multimedia in decision environments.

- Teleworking

  Perhaps the simplest example of a distributed work-group is the clerical teleworker, who works at home while accessing a remote system over a data link and interacting with colleagues over a videophone. This subject is covered in Chapter 9, which describes the well-known 'Inverness Experiment', involving ten BT Directory Assistance operators working from home, but with network connections to the normal Directory Assistance Bureau. One of the major discoveries from this experiment was evidence to support the hypothesis that teleworkers benefit from a rich support environment providing formal and informal contact with colleagues and managers.

  The operators were provided with video and voice-only telephony, e-mail and electronic whiteboards. Video was provided over basic rate ISDN. Each of the facilities was used several times a day, the videophone most frequently. Formal psychological testing demonstrated that the environment was indeed highly supportive. The value of videotelephony was graphically illustrated when its temporary withdrawal, for operational reasons, was reflected in a marked reduction in the overall acceptability of the system! A further important conclusion drawn from this and other incidents, is the critical need for reliability and good maintenance procedures.

- Desktop multimedia

  The teleworking experiment is an example of a very simple distributed working process, in that the individual operators perform tasks that are not in the main coupled to the tasks of the others and operate as independent clients of a main computer. Many tasks involve a greater degree of co-operative, peer-to-peer working, and many of these would benefit from improved human judgement through the collective use of multimedia.

  One such support tool, recently made available by advances in technology, is the desktop multimedia terminal. Chapter 10 addresses the components necessary for producing desktop equipment and the associated control systems, that make multimedia communications a genuine solution to distributed working. Desktop multimedia is much more than just a technology for connecting personal computers to the telephone network to provide simultaneous transmission of video, speech and data. If it is to be genuinely useful, it must provide full telephony functionality, including basic functions such as operation in power-down mode, as well as boss/secretary working.

Also, it must interwork with other units within the company and with other users, world-wide, not just through telecommunications but also office automation. Standards clearly are an issue, and a comprehensive account is given of the various media coding and aggregation standards, and those addressing network control and applications. The standardization process has been highly effective and products based on H.324, the PSTN multimedia standard, are now emerging. T.120, which includes a multipoint capability, has now been fully ratified, with products available from a number of companies, thus bringing into reality, desktop multimedia conferencing. A good source of standards information can currently be found on the Internet [10], but it should be mentioned that ITU documents are better purchased direct from the ITU, as other sources tend to contain drafts, rather than the final issued versions.

1.5.10   Integration and convergence

As shown earlier, the principal issue that needs to be addressed is the relationship between the multimedia communications service and the total business process it supports. As with any business automation solution, ubiquity of delivery, interoperability with existing services and technologies, scalability, etc, are more important than high, point functionality.

The prime challenge for decision-supporting multimedia communications is its successful integration into the IT and telecommunications infrastructures that run the business processes.

This implies that our video calls, our picture, video and audio databases should all be addressable within the company- (or even enterprise-)wide IT and communications networks, should be transferable and accessible through these networks in a manner consistent with their non-multimedia data architectures and accessed and controlled by the users' workstations and telephony equipment, via user interfaces compatible with other non-multimedia facilities. In practice, this may mean sacrificing Hollywood standards of images.

Notice that both IT and telecommunications were identified. Multimedia within the business environment is a hybrid between human-to-human interaction and process-driven computing tasks. Its requirements are based on human senses, but it is mediated through computer protocols. Sometimes this can cause problems. Chapter 11 highlights one such area, the problem of overlong and variable delay in local and wide area networks. These delays cause few problems when computers are communicating with each other, but on networks with even modest degrees of load, they can completely destroy the perceived continuity of moving images or speech. We shall return to this issue when we talk about the Internet later in this chapter. See also Baker and Paradiso [11] for further discussion of desktop multimedia in office environments.

Successful integration of multimedia communications into the business process calls into question the continuing division between telephony and computing.

As a corollary to this tendency of multimedia to force integration between computing and telephony, and the requirement of business to preserve the facilities of both, the development of new services, hybrids of computing and telephony, such as multipoint teleconferencing and multiple terminal interworking, are likely to be seen. Achieving standards for these complex interworkings at the same time as leaving room for competitive edge, across the telecommunications and intensely competitive computing industries, will be crucial.

1.5.11   Judgment and fun — virtual reality for business

Notwithstanding the dominant need for a ubiquitous utility of service in the business environment, the benefits of rich and realistic media presentation should not be ignored. Chapter 12 addresses the problem that arises with conventional desktop video because the camera is not concentric with the viewing screen. This creates the illusion that the distant party does not want to look you straight in the eye, leading to an adverse reaction to this perceived body language. They discuss a number of solutions to the problem.

At the very least, providing an element of fun within an IT system leads to significantly greater rates of take-up [12]. The advantages are not necessarily just in their entertainment value; they can contribute both to understanding the content and to navigating through it. For example, because images can be used to represent logical, rather than physical worlds, things can be placed together that share a close relationship rather than those that are close simply by an accident of geography.

Chapter 13 describes a virtual reality 'technical centre', which provides a common support service for documents and other information, to a number of factories around the world. Through its authorship, readership, etc, as well as its main text, a document's structure gives valuable clues to the way the work is structured. Representing the features of documents by geometric co-ordinates, the project displays them as clusters of similarity, showing for example, the common user groups.

Representations of the human interaction in the virtual business can be extended to creating virtual offices which locate people on the basis of the project structure rather than physical constraints. Anyone working on the project can enter the space and see who else is available.

Also in Chapter 13 is a description of a virtual fashion enterprise, where text-based information, images of cloth, yarns, etc, are integrated into a process description of the end-to-end business, even including a virtual fashion show with a virtual model on a virtual catwalk, modelling the garment.

The above activity was one of the ancestors of MASSIVE, a collaborative visualization activity led by Greenhalgh at Nottingham University, in which BT is also involved. This project uses shared spaces and simplified representations of human shapes, together with interesting ways of indicating when you approach someone 'close enough' to converse with them. Further information can be obtained on the Internet [13]. In conversations with participants, the author has heard that one of the limitations has been the lack of good-quality speech transmission over the Internet connection, perhaps confirming earlier remarks about the need to pay more attention to the issues associated with the audio medium.

1.5.11.1   Virtual reality on the Internet

The business examples given in the previous section are all driven by a simple theme — a belief that knowledge and knowledge organizations possess an underlying structure that can be described in spatial terms and can be represented using simplified graphics. This strikes an immediate chord with the exponentially growing number of Internet providers and users, where the issue of 'information management' has become critical. Therefore it is not unexpected that the Internet community has developed a number of tools for data visualization, and some are described in Chapter 14.

Early Internet standards, using HyperText Markup Language (HTML), permitted display of data that was mainly passive and two-dimensional and simply a way of passing pictures and text files without much in the way of an underlying semantic model. In Chapter 14 it is argued that we are now entering into a stage whereby new developments allow us to achieve this through the easy creation of interactive visualization and virtual environments. The corner-stones of these developments are Virtual Reality Markup Language (VRML), proposed as the *de facto* language for interactive multiparticipant simulation, and programming languages such as Java, specifically designed for distributed computing environments.

VRML provides a way of describing dynamically changing, three-dimensional models, and will soon include the ability to support interpreted scripts written in languages such as Java. This will allow downloaded objects to be brought to life on the local host machine and interact strongly with the user.

Given that the technological advances in local computing power are still outpacing the provision of wideband delivery, the ability to run powerful local processes will enable a new generation of interactive simulations. For instance, one experiment at BT Laboratories [14] involves the local generation of a fully textured image of a human head, whose mouth and general facial expression are made to move by a low bit-rate stream of commands extracted from analysis of an image of a live person at the other end of the communication channel.

This is an extension of the way that the development of speech and video codecs for use over limited bandwidth has led to a better understanding of the underlying low-level models of speech and image [15, 16], and bridges the fuzzy divide between 'coding' and 'artificial intelligence'. We can expect a similar depth of understanding to emerge regarding deeper semantic structures that describe virtual businesses, the entities that comprise co-operative teams, the mental models that allow us to understand large volumes of data, etc, and their subsequent realization in the form of multimedia virtual worlds.

1.5.11.2   Virtual reality tool standards

We should not leave the subject of multimedia representation tools without commenting on the similarity of the aims of VRML and MHEG, and other offerings, in their attempts to provide a descriptive framework for multimedia — and their different origin; MHEG comes from a telecommunications, video and photographic environment, whereas VRML has its origins in computer graphics. Despite the name, VRML is not a close relative of the document mark-up languages such as SGML, HTML and their multimedia extension, HyTime [17], which offer yet another approach. To say that HyTime is based on a view of multimedia as just information embedded in a documentation structure — moving video clips inserted in windows in electronic pages like photographs in a newspaper — is as simplistic as to say MHEG is too preoccupied with serialization and synchronization and not enough with database management, or that VRML is only satisfactory at describing imaginary geometric shapes, but currently it is probably true to say that they all tend to be biased towards their own origins (telecommunications and video, graphics, document architectures) rather than complete solutions to multimedia problems. There is as yet no universally satisfactory, or universally accepted, toolkit for the manipulation of multimedia entities.

1.6   COMMUNICATIONS — OPTIONS AND ISSUES

We have already mentioned the trade-off between media richness and the ability to access it from a distance. These tensions have, in some cases, contributed to the general thrust for wider band, more flexible, communications networks. For instance, we note the advances in Asynchronous Transfer Mode (ATM). However, the general area of ATM and its specific role in transporting multimedia has been widely covered in other publications (see, for example, Adams [18]). We prefer to concentrate on other transmission systems which are particularly determined by the characteristics of the multimedia requirement.

1.6.1   Delivering entertainment services

Chapter 7 discusses asymmetric digital subscriber loop (ADSL), the major transmission breakthrough that may make video-on-demand realisable in the near term, over existing copper cables into the home. However, used alone, this would require servers to be installed within a few kilometres of every customer, which would be expensive. There is also the issue of how these servers could be downloaded with content. Some calculations indicate that this could be greater than the total capacity of the current telecommunications network. Thus, to make ADSL useful, it needs to be incorporated with other transmission technologies. One possible configuration is shown in Fig. 1.5.

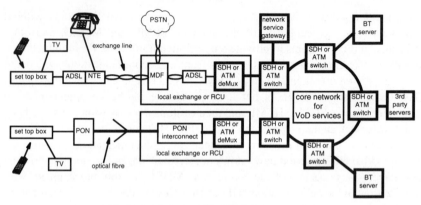

Fig. 1.5   Video-on-demand network architecture.

ADSL is still used for the last few kilometres, but is connected to an SDH or ATM demultiplexer. This multiplexer is in turn connected to an SDH or ATM switch, which is one of many on a wide-area SDH or ATM ring which also includes the media servers containing the content material to be accessed. As an alternative to ADSL for local delivery, fibre, radio or coaxial cable can obviously be fitted into this structure.

Some US cable companies believe that 'hybrid fibre/coaxial (HFC)' can offer a cost-effective solution, with fibre carrying the multichannel signals from the central office to a convenient distribution point capable of covering a few hundred to a few thousand customers. The final path between this distribution point is carried on coaxial cable, with each customer supplied via an individual cable. Digital signals are transmitted over this cable, in both directions, via RF 'cable modems', which can provide, typically, up to three bi-directional digital video channels [19]. An interesting debate surrounding this implementation is in the granularity of the number/bandwidth of channels to be provided. Should providers aim for wideband, high-quality video, or narrower band, multiple channels for voice telephony and Internet connection? This is also an issue for ADSL copper systems, which will only be resolved once the relative popularity of video-on-demand versus home Internet connections is determined.

Under wide-scale deployment, the structure of video-on-demand may be more complex, with central servers containing information for very wide area distribution, 'metropolitan servers' covering, as the name implies, major conurbations, and local servers, providing frequently accessed or local information. See Furht *et al* [20] for a detailed description, albeit predominantly from a USA perspective, of how telecommunications and cable networks will evolve to this model.

There still remains a lot to be understood in the dimensioning of multimedia servers and their associated networks.

Bi-directional, high-speed transmission to the home is also emerging in the form of 'HDSL' [21]. In principle, this would provide a high-quality interactive service for home-working or mass-market videophone.

One area requiring further development is in the best choice of distribution medium for multimedia services within the home. There is no clear leader in the home distribution market, either in terms of transmission medium (coaxial, cable pair, radio), protocol (ADSL, ATM, LAN, e.g. token ring), or vendor paradigm (computer, telecommunications or consumer electronics). To a large extent, this will be determined by the winning structure for the home multimedia terminal — interactive TV set-top box, or home computer, in particular.

1.6.2   Transmission techniques for business multimedia and kiosks

Unlike entertainment networks, the requirements for distributed working depend less on very wideband (though predominantly uni-directional) provision and more on bi-directional, low-delay, ubiquity of service. To a lesser extent, this is also true of full-function kiosks. In the short term at least, this is a significant opportunity for ISDN, certainly at basic rate and perhaps with aggregated channels of six or even more.

However, on its own, this is not much more than a videoconferencing technique, or, at most, a solution to the interconnection of executive desktop terminals connected to direct exchange lines. As explained above, multimedia communications also needs to be integrated into the IT infrastructure. This needs a rethink to LAN architectures, in order to provide guaranteed bandwidth and low delay. Given that ATM may be the natural evolution path for distributed working multimedia over the wide area, it will be interesting to see how far it penetrates intra-building standards. There are currently integration issues with 'telecommunications' designs such as ATM and 'computing' technology such as TCP/IP, owing to the former's fixed packet length.

1.6.3   Interactive multimedia communication on the Internet

The problem of bandwidth and delay is an issue which is exacerbated when we consider Internet services [22]. Quality of service is not defined for the Internet. Since it consists of a set of loosely co-operative but autonomous computer sites, whose wide area connections usually involve a number of intermediating routers which can assign (and alter) network paths on a packet-by-packet basis, then it is not surprising that route competition and network congestion lead to highly variable transit times and packet re-ordering. The net result is the introduction of significantly long and unpredictable end-to-end delay and restricted bandwidth.

Until a fully guaranteed solution becomes available, a number of experimental 'fixes' are being pursued. A particularly interesting example is the Multicast Backbone, or 'Mbone', a virtual network intended for multicasting audio and moving image, including video conferencing, over the Internet. It recognizes that, given sensible controls on packet routeing and packet life, it is not too destructive of network capacity to multicast data to a number of sites. An on-line description of Mbone is available on the Internet [23].

One realization of Mbone, in common with a number of other multimedia solutions, makes use of an experimental protocol, RSVP (resource reservation set-up protocol), which lets users request bandwidth or other network resources along a path of routers. This allows the setting up of a more-or-less guaranteed 'connection'. Some authorities believe that this is unduly restrictive and suggest that 'fair queuing' algorithms, or models based on real-time escalation of the price of service, are more likely to preserve the rapid and robust growth of the Internet. Mentioning price reminds us that this is very much a business area, at the mercy of the market-place. A good summary of the present industry position is given in Hart [24].

As with other solutions, multimedia on the Internet tends to be associated with 'image', either moving or still. However, it is important to note that the provision of adequate audio is at least as difficult an issue. In the case of interactive services such as conferencing, the problems of coding delay and, even more so, retransmission in the case of packet error, pose severe problems. Even where audio is not 'two-way', for instance in on-line access to entertainment services, delay and error are a bigger problem with audio than video, because the way we perceive images as opposed to sound, particularly speech, means that error concealment is much easier with image. The reader is referred to Wheddon and Linggard [15] for details of speech coding techniques.

1.6.4   Multimedia on the move

The subject of multimedia communications techniques would not be complete without mentioning two successful applications which required wireless communication on the move. There almost seems something perverse in the way BT has tackled the issue of providing multimedia content-based services — the first offerings have been (almost literally) launched from aircraft and from yachts at sea!

The in-flight system for information, shopping and entertainment services has already been mentioned. This consists of a local multimedia platform located on the aircraft, with a (non-multimedia) air-to-ground (direct or via satellite) link to carry data traffic, such as the purchase orders for goods or the despatching of facsimile messages. Use of the data channel, which operates over the Skyphone/ Jetphone service [25], converts the in-flight service from a browser to a full-function terminal.

Compared with the stability of an airborne platform, bringing back pictures from yachts at sea presents an even greater challenge. Chapter 15 describes how moving pictures and sound, to a quality acceptable for TV broadcasts, were composed on-board and then transmitted at a low bit rate (64 kbit/s) reliably over a satellite link, despite the extremely adverse transmitting conditions, using a novel error-correction system.

1.7   CONCLUSION — CONVERGING ON THE SUPERHIGHWAY

So far it has been shown that the difficulty in defining multimedia communications has come about by the varying and significantly different human requirements being met by tactical selections of technologies chosen to meet only the key human requirements. This impression will be reinforced by the more detailed chapters that follow.

We see that content-driven services have chosen different transmission, coding and terminal systems from those of interactive, co-operative working. We see a significant divide between consumer electronics, telecommunications and computer industries, in the same areas. Is the home platform of the future to be a TV set-top box or a personal computer? Is the network architecture of the future a telecommunications one based on ISDN/ADSL/ATM paradigms or a computer one based on Internet? If we are to achieve fully the requirements of multimedia, the answer is probably, 'all of them — and perhaps a bit more'. Resolving these issues and achieving harmonization at the technology level, while leaving room for the development of a wide choice of products, is the domain of the so-called 'convergence' debate. This debate is complex and not yet fully defined. This book intends to contribute constructively to the debate.

When, and not before, these issues are resolved, then and only then, will the 'Information Superhighway' really come into being, providing the ubiquitous delivery of entertainment, information and instantaneous, bi-directional, media-rich communications. At that time, we will at last easily understand what is meant by 'multimedia communications'.

REFERENCES

1. Williams N and Blair G S: 'Distributed multimedia applications: a review', Computer Communications, 17, No 2 (February 1994).

2. ICCC Conference, Keynote address: 'Multimedia Communications 93', Banff (April 1993).

3. Stephan O: 'Digital video spearheads TV and videoconferencing applications', Computer Design (December 1994).

4. Ristelhueber R: 'MPEG is the key to the info-tainment explosion', Electronic Business Buyer (July 1994).

5. Hammer D: 'Standards caught in multimedia cross fire', America's Network (January 1995).

6. 'The gathering storm in high density compact disks', IEEE Spectrum (August 1995).

7. http://java.sun.com/about.html

8. http://www.genmagic.com/Telescript

9. Anderson C: 'The accidental superhighway', Economist (1 July 1995).

10. http://www.csn.net/imtc/

11. Baker R and Paradiso T: 'Integrating video at the desktop', Telecommunications (December 1994).

12. Igbaria M, Schiffman S and Wieckowski T: 'The respective roles of perceived usefulness and perceived fun in the acceptance of microcomputer technology', Behaviour and Information Technology, 13, No 6 (1994).

13. http://www.crg.cs.nott.ac.uk/~cmg/massive.html

14. http://transend.labs.bt.com/head/about/maxis.html.

15. Wheddon C and Linggard R (Eds): 'Speech and Language Processing', Chapman & Hall (1990).

16. Gonzales R C and Woods R E: 'Digital image processing', Addison-Wesley (1994).

17. DeRose S J and Durand D G: 'Making Hypermedia Work (A user's Guide to HyTime)', Kluwer Academic Publishers (1994).

18. Adams J L (Ed): 'ATM for Service Providers', Chapman & Hall (to be published in 1997).

19. Anderson V: 'A match made in heaven', America's Network (October 1994).

20. Furht B, Kalra D, Kitson F, Rodriguez A and Wall W: 'Design issues for interactive television systems', Computer (May 1995).

21. Baker G: 'High bit rate digital subscriber lines', IEE Electronics and Communications Journal (October 1993).

22. Crowcroft J: 'The Internet: a tutorial', Electronics and Communication Engineering Journal, 8, No 3, pp 113-122 (June 1996).

23. ftp://taurus.cs.nps.navy.mil/pub/mbmg/mbone.html

24. Hart K: 'Building the global Internet backbone', Communications Week International, pp 17-23 (July 1996).

25. Stuart A: 'Skyphone', British Telecommunications Eng J, 10, Pt 4, p 336 (January 1992).

2

DESIGN OF MULTIMEDIA SERVICES

G Smith and W Jones

2.1 INTRODUCTION

2.1.1 Initial concepts

Before a 'service' can be designed it must be conceived. There is fundamental commercial and technological research to be done before embarking on the long journey of designing a multimedia service and the applications which live therein. There is an order which must be obeyed to achieve success and a structure to be designed and built before it can be decorated.

High technology projects are often late, over budget and much more complex than first thought. There are many examples in the world to substantiate this statement. This does not have to be the case but how do we ensure success? Though this chapter cannot answer all these questions, it attempts to identify and relate the issues and tasks which have to be integrated to create a multimedia service and to populate that service with well-designed and timely applications. Experience with previous multimedia projects is now building and there are many lessons to be learnt.

Technology-based companies like BT often perceive multimedia services as an exercise in applying incremental improvements in technology, building on what has been before. The very small amount of successful multimedia world-wide is generally confined to non-networked personal products such as CD-ROMs. There is no model for the successful design and implementation of net-worked multimedia services, though this may be remedied through the many world-wide video-on-demand (VoD) trials being planned and implemented. This chapter divides the creation of a multimedia service into two parts.

Service design — which embodies the commercial rationale, business opportunity and technology exploitation. It also includes the scope and scale of the service and defines the service hardware and software platform. The outline requirements for the applications must be established in aspirational terms to drive the platform functionality and choice. The platform functional specification and user interface (UI) specification are established at this stage, thus providing a generic environment for later application creation. Application creation is not started during this phase.

Application creation — the design and creation of the applications which comprise the service. A function of an application is to present a user interface to the user which provides navigation and selection of multimedia content. Creative design, usability and content, within the platform technical constraints, are the key features of a multimedia application.

2.2   SERVICE DESIGN

The service design process is described in this section and illustrated in Fig. 2.1.

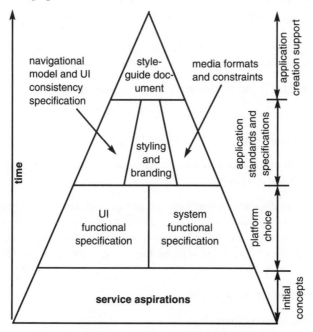

Fig. 2.1   Service design pyramid.

Figure 2.1 is pertinent to the entire section and the subsections which describe and explain the elements in detail. The time arrow indicates the correct sequence of work. The pyramid shape shows the cumulative build-up of knowledge and the interdependency of the elements, cumulating in the production of the style-guide document, which provides the basis for application development.

2.2.1   Service aspirations

To drive the service design forwards the basic aspirations for the service must be established. There are two components:

- short or target list of business strands to be offered by the service, a business strand being defined as a generic commercial area within the service, e.g. shopping;

- initial concepts of how each business strand can be represented by interactive multimedia technology.

A requirements capture exercise is generally necessary to establish the generic requirements of each business strand. This could be done with a selected 'information provider' partner for each area, who would collaborate in the development of a 'lead' application, and would be offered an inaugural place in the service line up. Alternatively a non-aligned source can be sought.

The requirements capture will take the form of a number of workshops whose output will be aspirational specification documents covering:

- typical ranges of products to be offered by each business strand;

- product categories and themes for selection and searching;

- typical total amount of product to be offered;

- the mix of media to provide information about the products, for user navigation and selection;

- choice of media used to represent the product contents — for entertainment services the 'product' is data transmitted to the user, e.g. a movie or TV programme, while for other business strands, such as retailing, catalogue stills and short video clips will be used to provide information about the product;

- business transactions required;

- usage statistics required;

- UI and user interaction.

2.2.2   Choosing the platform

The aspirations for the service, described in section 2.2.1, are to some extent preconditioned by a knowledge of the current state of multimedia technology. The UI and user interaction options available are tested and balanced against the aspirations for product representation and display. The ultimate choice of platform will be a design compromise.

The factors for consideration in choosing the platform are:

- display technology — TV, computer monitor, LCD, etc;

- display attributes — size, resolution, visibility, siting and ambient lighting;

- UI interaction device — TV-style remote control unit, touchscreen, mouse;

- media storage and distribution — the total on-line media requirement for the service and the distribution of its storage between a central server and local storage device;

- media types — the different media types to be used by the service (video, stills, text, audio), the critical media type being video because of its high storage requirement and impact on network bandwidth;

- performance — the service performance requirements for media display and user interaction feedback will be quantified as maximum response times for all functions within the service, e.g. time to display a video;

- cost — the functionality and performance of the platform are directly related to its cost, the final specification always reflecting a trade-off;

- connection to external systems — requirements for business transaction or networking to external servers.

This chapter looks at generic rather than particular multimedia platform solutions and a generalized platform architecture is depicted in Fig. 2.2.

This generic architectural model is applicable to most particular platform solutions. Though the relative importance of the central components, the network server and the central media store may vary, the model holds good as a starting point for the design of new service platforms.

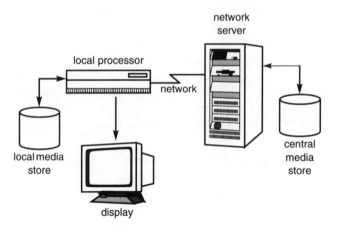

Fig. 2.2   Generic multimedia platform architecture.

There are five principal components of the generic multimedia platform.

- Local processor — the local processor executes the user interface code, driving the main display unit and peripherals and providing an interactive capability for user interactions. The local processor provides the local component of the service processing power. The functionality of the local processor is dependent on the distribution of the application media contents between the central and local stores and also the performance criteria for finding and displaying the media content. For some 'stand-alone' kiosk services there is no network connection and a single processor supports the complete service. For others, such as VoD, the local processor provides the UI, calling central services over the network, when required. The UI also provides the means of displaying central data on the display, for example, playing movies transmitted from the central media store.

- Local media store — the local media store can provide the media content displayed by the application. This comprises still images, text, sound and video. Locally stored video is generally confined to shorts clips because of the high storage requirement. In a kiosk service, typically, there would not be more than 30-60 min of stored video. If stored on a hard disk the stored video can be swapped out, by overnight 'store-and-forward' techniques, from the central media store. With VoD, there is no local media store, all the application content being held on the central media store-and-supplied to the UI, on demand.

- Central media store — the central media store generally provides a centrally networked source of media content required by the application, comprising still images, text, sound and video. Centrally stored items are generally longer videos and even feature films in the case of VoD services. The central media store can have two functions depending on the service requirements and network bandwidth:

  — to store and forward media to a local processor unit, either 'on demand' or as part of an 'off-line' contents maintenance strategy;

  — to supply media content 'on demand' to local processor units for real time display.

- Network server — the network server provides functions which are executed by a client/server call from a local processor unit. A function could relate to content, for example, finding and transmitting a video clip, or to a business transaction. In the case of a VoD service the central server is all important and can comprise a number of linked devices providing the UI, the application media contents and the business transaction capability in a multi-user, real-time environment. For kiosk services the network server provides text database retrieval on demand, for example, retail product prices and availability. It could also provide real-time still images if they are stored in a suitably compressed form, such as JPEG. If the network link were broadband, real-time uncompressed still images and compressed video could be provided. Broadband provision is unlikely for kiosk services. For this type of application, the media contents could be held in the local media store which would be replenished off-line by the network server.

- Network bandwidth — network bandwidth is a trade-off between the aspirational media and performance requirements for the service and the installation time-scales and operating costs of network technology. For VoD services the exploitation of broadband technology is an explicit part of the service concept. For other services the 'ubiquity' of the network is an important factor in the time-scales for exploitation and is intimately connected with the strategy for media distribution and storage, since performance is critical to service usability and user acceptance.

2.2.3   Platform functional specification

The platform functional specification defines the functional behaviour of the platform. It translates the service requirements into functions which form the application building blocks. The platform functional specification is expressed at a high level but with sufficient detail to enable a software developer to produce a technical specification and write the computer code. The scope of the functions covers:

- product data search and retrieval;

- media retrieval (video, stills, audio);

- business transactions;

- usage statistics;

- gateways to other systems.

The aspirational requirements for the service, as captured and documented through requirements capture workshops, are often not fully supported by the capability of the platform and some downgrading of expectations will occur. Service design has to manage compromise so as to maintain acceptable levels of service viability, despite platform constraints.

2.2.4   UI functional specification

The UI functional specification defines the building blocks which provide the application designer with the tools to design the application UI. It defines the user interaction paradigms which are the routes to application content searching, navigation and media display.

These interaction paradigms, using the previously defined interaction devices, are specified in detail. The UI specification assigns and maps user action to functionality in a generic, non-application-specific manner, setting out generic rules for later incorporation in application-design standards. The UI specification covers:

- selection methods for 'product' search and UI interactions to specify search criteria;

- display standards for the media returned following a product search — the mix of media types is defined here and, in particular, the control and access to video material are specified, examples being the standards for continuous video loops and video carousels;

- selection of on-screen objects and feedback;

- UI components of service-specific functions, such as 'shopping basket' for retail services and 'movie pay-and-play' for entertainment services.

2.2.5   The navigation model and UI consistency

The navigation model shapes the generic UI specification one step nearer the applications, providing one or more navigational templates which when tailored for specific applications, provide the core, search, navigation and display engines for those applications. The navigation models are the integration of the components defined by the UI functional specification and are a higher level concept. At this stage, fundamental navigation conventions are agreed to provide consistent ways of implementing basic functionality throughout the service, e.g. a consistent way of exiting all applications. This may simply mean adopting only one possibility, as defined in the UI specification, or designing a more complex structure from UI building blocks.

A common thread of UI consistency, once defined, will be built into all navigation models, to give a consistent core 'feel' throughout the service. The 'look' will and should vary throughout the applications, to reflect the aspirations of the service providers.

The navigation models are essential components of the service design process and are themselves essential components of the style-guide document, described in section 2.8.

2.2.6   Media specification, formats and constraints

Media specification, formats and constraints are major components of the design of a multimedia service, influencing media contents production, media storage on the system and performance of the service with respect to media retrieval and display.

Some basic media standards are fixed by the platform choice while others will offer some flexibility and choice, allowing final decisions to be made as part of the application development process, discussed in section 2.3.1. In particular, video standards are critically important, as video is the most resource-hungry of all the media, being exceptionally difficult to produce, encode, store and display, in comparison with other media types.

A media specification, format and constraints document will be produced at this time. This will progress the standards explicitly and implicitly emanating from the hardware platform, and will form the basis of the media section in the style-guide document which will be used by the application development teams. This document will, in general, be expressed in technical terms and will form the definitive technical reference for the service. It will be available to application design teams which will use it for background reference, using the style-guide as their primary reference for design information and support. The document will express media and performance constraints in purely technical terms and will not attempt to develop design guidelines or advice, which are solely the province of the style-guide document. This document is a major input into the style-guide document, described in section 2.2.8.

2.2.7   Overall branding and styling

The overall branding and styling represent the fundamental values and character of the service and must be in position as early as possible so that they can be included in the style-guide and hence available to application developers before the commencement of application development.

The ownership of the service and hence the assignment of branding are fundamental decisions to be made. For a simple service with a single service provider using their own brand throughout there is no problem. For more complex services, for example VoD services, with many information providers, the situation is much more complex. In this case the overall service provider must decide ownership of the 'top levels' of the service and the use of their own branding in these areas. The transition to the branding of the information providers will often require negotiation and the information providers will have to 'buy into' the overall standards and values of the service, otherwise it will be difficult to sustain the long-term commercial partnership which is generally desired by both parties. For the above reasons it is imperative that the branding and overall design standards are put in place before the applications are started, otherwise many misunderstandings and mistakes will occur involving loss of time, credibility and expensive reworking of designs and media content.

The place to document overall branding and styling is the style-guide which is described in the next section. However, examples of the 'top-level' design work would generally be available as part of an on-going consultancy exercise and efforts would be made to involve and consult information provider partners. However, it is the service owner who ultimately sets the rules for itself at the top levels of the services and, by negotiation, in other branded areas of the service, i.e. the applications.

2.2.8   Style-guide document

The application style-guide is arguably the most critical and important document produced by the service design process. It documents the service design standards, as described in sections 2.2.5-2.2.7, to provide the definitive reference manual for application design and media contents production.

It is important to understand the significance of the main inputs to the style-guide which are repeated below:

- the navigation model and UI consistency;

- media specification, formats and constraints;

- overall branding and styling.

The style-guide interprets, where necessary, the service features and constraints to provide mandatory guidelines and offers design alternatives where possible.

The style-guide document cannot be written unless the three principal inputs exist. It is then possible for an experienced interaction designer to produce the document. However, it is not recommended that this approach is taken. The best approach is to have a fully integrated service design team from the start, involved in all aspects of service design, especially the three direct inputs to the style-guide. The style-guide document will then grow organically from the service design process, being fully available and supportable at the right time.

The style-guide is a 'living' document which is fully supported as part of the application creation process, which is the next major section of this chapter.

2.3   APPLICATION DESIGN

2.3.1   Process model

In order to be able to design an application, a process model is required. The following model shows the different activities that take place during the early stages of application creation, up to and including the definition and sign-off of the application design document. Requirements capture techniques (see section 2.3.3) are needed to capture the requirements of customers, in early meetings and workshops with the customer. These requirements should not be constrained to the functionality of the technology. Customers should be allowed to express their aspirations for the application and service. These aspirations can be used to enhance the technology, during future upgrades. Included in this process, is the information on corporate styling and branding. Requirements and understanding of the brand values the customer wishes to express within the application and service will be captured and understood.

This information, and the information included in the style-guide (see section 2.3.2), allows the creative designer or interaction designer to work out the interaction and navigation of the application. These are drawn up as visual storyboards (see section 2.3.3). These storyboards are reviewed with the customer at regular workshops. Once agreed these are worked into concept screen designs. These screen designs show how the application would look once complete. They would follow all the consistency requirements of the style-guide. Again, on

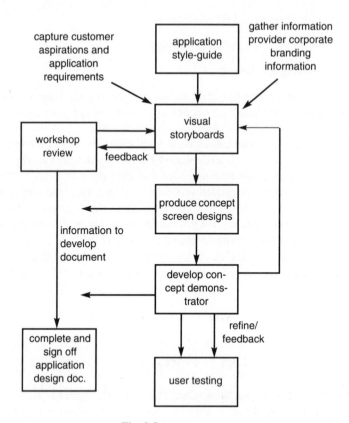

Fig. 2.3   Application process model.

agreement, these are put together by a prototyper, using the visual storyboards, and a prototyping authoring package and turned into a concept demonstrator.

This demonstrator is tested (see section 2.3.3) with the target users. Human factors facilitators are able to explore and test the usability of the application with the users. This valuable information is fed back into the design process. Where required, the storyboards and the demonstrator are altered to reflect the user testing recommendations.

During this whole process, information is being gathered for the application design document, sometimes called the design brief. This is the definitive document for the design of the application, and must be agreed and signed by the customer.

All the activities described, and shown in the model (see Fig. 2.3) must be co-ordinated. Therefore an overall application design manager is required at least until completion and sign-off of the application design document.

2.3.2   Design support

2.3.2.1   Style-guide

The style-guide is the 'applications bible'. It is used by the application design team, during development of the visual and technical storyboards, during early prototyping and the production of the application design document. Its purpose is fivefold:

- to document the navigation model of the service, from the UI functional specification developed through the service design process, thus ensuring that the user interaction and navigation is consistent in all generic aspects of each application, i.e. so that all applications work in the same way — for example, during payment, the order in which the user is expected to enter data would be consistent, thus reducing the learning task and minimizing confusion within the different applications;

- to prescribe, in detail, core areas of the user interface consistency, e.g. ensuring that the user can always expect help, information or error messages to be consistent in both look and style — interface consistency, however, should not restrict the styling and branding of a third-party information provider;

- to state mandatory core standards where applicable — these may apply for designing on the target screen displays, or for the platform;

- to document all user interface constraints imposed by the platform;

- to promote best practice, within the 'art of the possible', in user interface design, e.g. the amount of text to be used on screen, and the minimum size of the text that should be on screen.

In addition, the style-guide must cover the following topics in detail.

- Styling and branding

    This section of the style-guide needs to clearly document the positioning of the brand and brand values of the service provider. These decisions will have been made during the service design stage, before application design commences. The brand and its positioning will need to be reflected in a branding system that will be imposed on to the application design. The branding system will reflect the brand of the service provider, the brand of a third-party information provider and the transition between the two.

The branding system would define the attributes that reflect the brand — the use of corporate fonts, the positioning of service and application titles, the position of corporate and third-party brands or logos, the use and positioning of sponsorship logos, and the transition between corporate and third-party corporate branding.

The BT market trial branding system, for example, carries the feel of the BT corporate brand through to the main screen of each application. Each application main menu has a branding strip on the left hand side of the screen. This strip contains the BT brand, the icon that represents the application area and the title of the application.

- User interface consistency

    The user interface should cover what the user sees on the display screen, the interaction devices used to interact with the screen, and any literature that may accompany the service.

    The following are the sections that can be expected to be covered in this area of the style-guide.

    User interface display — information on the type of display to be used, is provided. For example a computer screen, a TV screen or an LCD display. To design for a particular display, the designer needs information on the size and resolution of that display. Environmental information is also specified, such as information on ambient lighting, reflections, and the viewing distance of the user from a screen.

    Interaction device — this section gives full details on the device that will be used to interact with the application. This may be a mouse, touchscreen, or infra-red remote control. Each device has limitations which affects the design of the user interface, for example, a touchscreen would have a minimum setting for screen objects.

    Colour — colour display is dependent on the display hardware being used. For example, a TV in the UK can only truly display PAL-friendly colours — colours outside of this range will badly bleed across the screen or flicker. Technical standards are available, and should be quoted. Human factors standards provide information on how many colours should be used on screen at one time and describe the emotional significance of certain colours, e.g. red is representative of stop or danger.

    Text — this section recommends usage of fonts within an application. These recommendations include text sizes, font types and text presentation. Minimum font sizes are required for different text. This text could be a title, product description or price information. Font sizes are dependent on the viewing distance of the user to the display. Some text will not change through the lifetime of the application. This 'static' text is anti-aliased to the background image. The font used is normally a corporate font, or one that is best displayed on the target display. However, some information must be updated regularly, for example, news information. This information is stored in databases, and cannot be anti-aliased, giving it a less smooth look. This 'dynamic' text is represented in a default system font, typically Helvetica. Presentation of text is available from human factors standards, and covers the tone of the text or phrase, where it should be displayed, and how it should be displayed.

Sound — sound can be represented in three ways — audio feedback, background music and voice-over. This section details the use of music and voice-over, especially when it should be used and how frequently. The tone and tempo of the music should reflect the desired atmosphere sought for the application. The music should not be too repetitive, this is annoying to the user. The same applies to voice-overs. An appropriate voice should be chosen for the application. Voice-overs should support the text used on screen.

Imagery — the way in which imagery is presented on screen is dependent on the display hardware and the size of the display. Recommendations need to be made for background and foreground static images as well as video content.

Selection — selection can be direct or indirect. Direct selection occurs when a single interaction causes an action. For example, a choice on screen is associated with the number '1'; by pressing '1' on a remote control the user would automatically be taken to a new screen representing that choice. On interaction, both audio and visual feedback is given. Indirect selection requires two interactions from a user. This first is to move a 'focus' visual highlight from one option to another, the second interaction is to select the focused option. Two different visual highlights are required for this method. A user should be able to recognize the different selection method and interact accordingly. If the selection choice is small, the direct method is generally preferred.

Feedback — feedback can be tactile, audible and visual. The user will get tactile feedback as they press a button on a remote control, a visual feedback will be displayed on the screen, and an audio feedback can be heard.

Messaging — this section gives recommendations for global system-error and information messages, and local application-error and information messages. These recommendations will cover how the messages will be displayed, to draw attention to the user, and the wording associated with the message. Messages are normally represented in pop-up boxes that appear over the screen the user is viewing. Messaging should be consistent throughout the application and indeed across the full service.

Help — there are three types of 'Help' — tutorial help, a hierarchical help system, or context-sensitive help. All three should be used if the platform supports them. The first could be, for example, a guided tour around the application explaining the fundamental user interactions. Hierarchical help has full search methods to find out any information on the application. Context-sensitive help, is generally available for each screen, and will only offer help specific to the information on the screen. This help is often activated through a dedicated 'Help' key. 'Help' should be consistent throughout the application and indeed across the full service.

- Navigational consistency

Payment — the way in which a user may pay for a service or purchase a product must be consistent throughout all the applications. The user will require the mechanism to enter their payment card number, and expiry date, before then agreeing to the transaction. Security precautions may be required, in a form of a PIN.

Exit — the user must be able to exit an application and service in a consistent way. The method of exit and the information provided must be consistent, from wherever they leave the service.

Time-outs — time-outs are used throughout applications. Screens may 'time-out' and return to a previous point in the application, play a video after a certain time of inactivity, or automatically exit the user from the service, if the time elapsed for inactivity is exceeded. The types of time-outs within the service must be documented. The service time-out must be consistent for all applications.

2.3.2.2   Media visualization

It is essential that early visualization of the screen designs are viewed on the target display as soon as possible. The target hardware can vary greatly from the development environment. For example, most graphic development is done using a computer, but the final graphics might be displayed on an LCD or TV screen. These displays can vary the contrast, brightness and luminance of the graphic image, as well as allowing colours to 'bleed' or flicker.

2.3.3   Functional elements of the process model

2.3.3.1   Requirements capture

General requirements capture will have been carried out during service design. These requirements will document all the aspirations of the typical potential customers. Not all of these will be realized on the platform.

Requirements capture during application design focuses more on the possible, and covers in detail the navigation and contents to be designed into the application. Through these detailed workshops with the customer, information can be gathered and detailed into a requirements document. The document will contain:

- a high-level application definition, describing the application;

- detailed information on the navigation, and the number of high-level options or areas within the application;

- information on the scope of content to be presented in the application;

- tasks to be performed — purpose of task, user input to task, task outputs, task frequency, task goal, task dependencies, task duration, task flexibility.

Even, at this stage, some aspirations will not necessarily be met by the hardware or software. Changes can be requested to enhance the functionality of the platform to accommodate the customer's requirements.

These requirements are then taken a step further and developed into storyboards.

2.3.3.2   Storyboarding

Storyboarding is a very useful technique for visualization of the structure and navigation of a multimedia application, following the requirements capture stage.

The technique was first used in animation, but is now widely adopted for multimedia application design.

Storyboards allow the interaction and graphic designer to lay out and design an application in visual terms. Quick sketches can be linked together, allowing the application to grow, and for different navigation flows and application ideas to be visualized.

Storyboarding is successful because at an early stage it is very flexible, allowing the designer to sketch out many different ideas. Consequently, this technique is used throughout the early stages of application design. The three types are:

- concept storyboards;

- visual storyboards;

- technical storyboards.

Concept storyboards can be as simple as rough sketches on 'post-it' notes on a large wall. The interaction designer can sketch out the screen images on individual notes. This technique allows the designer a chance to explore the navigation flow of an application, being completely flexible in moving the 'post-it' notes or 'screens' around. This process can be done, with little drawing experience. Although the sketches need to be recognizable, text labels can be used to remind the designer of the screen image. At this stage of design, the storyboard is an excellent brainstorming tool.

This whole process allows the designer to map out the navigation hierarchy of the application and the individual tasks executed within the application, from the start of the application right through to the end-product presentation, i.e. define the application structure.

Once a certain design has been agreed, using simple flow-chart documentation, the concepts and application tasks can be fleshed out further in the form of visual storyboards.

Visual storyboards are also created early on in the process of application design. Application flow or navigation will normally have been agreed by this stage, but sometimes it is quite usual to visualize alternate design ideas.

Usually a 'slice' through the application is visualized. A little more care is taken with the storyboard sketches. At this stage thought is also given to application consistency and the possible layout of the screens, e.g. where the titling or the navigation buttons might be placed. Consideration is given to the exact interaction the user would make on screen.

This visualization can be done on paper or computer. Multimedia developers with Apple Macintoshes might consider using HyperCard. Individual cards can represent individual screens.

Concept storyboards represent the hierarchical and sequential nature of an application. Though a multimedia application may have a hierarchical structure, generally there is more than one route through the application. Applications tend to have numerous dynamic links, allowing the user to jump from one place in the hierarchy to another. For example, browsing though places of interest in a tourist resort, a user might want to book tickets to a theatre. This might take the user from the 'What's on' strand to the 'Reservations' strand of the application. This linking must be recorded in visual storyboards.

Visual storyboards are often supported by early screen prototypes. These aid visualization for the customer who can, using the visual storyboards, map them to screen prototypes to imagine the end application. Where possible, the prototype screen designs should be viewed on the delivery platform, e.g. TV screen,

LCD display or a computer screen. This ensures that the customer has full understanding and visibility of the application at a very early stage of design.

This level of storyboarding still allows a great level of flexibility and is very cost effective.

Following the visual storyboard stage, these ideas are further tested by turning the storyboards into a concept demonstrator.

Technical storyboards will only be briefly described. This stage is usually completed following the sign-off of the design brief by the customer.

These storyboards will be produced for every single navigational screen and provide template information for large amounts of product data information. This storyboard gives full detail of every possible navigation to and from the screen, provides complete references to all graphics, sound files, or video material that make up a complete screen. As well as references, exact screen co-ordinates are also supplied. This allows the 'application' builder to take all source material and build the application with ease.

2.3.3.3   Concept demonstrator

A concept demonstrator is normally a 'stand-alone' demonstrator, that mimics the target platform as much as possible. Graphic designers can work with the visual storyboards, to provide full-screen graphics for the prototyper to build into an application.

The proposed navigational structure is put together, using a commercially available authoring package, and populated by the screen graphics. Two authoring packages that are frequently used to build demonstrators are 'Authorware' and 'Supercard'. Both are very easy to use, and a prototyper can quickly build a demonstrator.

This demonstrator can fully represent the different application strands. Only a subset of product media content is required for the demonstrator to be effective. A fully populated demonstrator ensures that it can be thoroughly user tested. A demonstrator that only has one 'route' through it may not fully reflect the application design; it means that only certain aspects of the application design can be tested. In certain circumstances this may be acceptable.

Having established a navigational model during the service design stage, user interface consistency and branding will be implemented into the demonstrator. The effectiveness of this model will be evaluated by subsequent user testing.

- User interface consistency

    Screen layout is a key to achieving UI consistency. It ensures that the user can navigate around the application with ease, without losing their sense of place in the application hierarchy. Consistency is achieved by designing templates. These template screens would have specific layouts, specifying text areas for titling, body text, possible pricing information, or image positioning. Using these templates for certain screens promotes a consistent 'look'. For example, a product detail screen would have a title of the product at the top, a description on the left, a product image on the right, and possibly a price below the product description text.

UI consistency, however, does more than suggest layout consistency. It refers to consistent use of colour, text sizes and fonts. Therefore, a product detail screen, would not only specify certain areas for text or graphics, but would also reference font type, font size, font colour and number of colours used for the graphic. All these interface details are reflected in the demonstrator.

- Branding

  Branding is essential to communicate the brand values of the application provider to the users. The application provider may be BT or indeed an information provider. The brand values expressed by BT, are those which it wishes to ascribe to the application. This is achieved through implementation of the BT branding system, thus differentiating our applications from other information providers and our service from those of our competitors. BT's brand values and branding system would be fully defined in the application style-guide.

  If the service is to take on a new branding approach then this needs to be designed and implemented using a new branding system. This can be evaluated during user testing.

2.3.3.4   User testing

User testing takes place at two different levels — expert evaluation and end user testing.

Expert evaluation is carried out usually by a human factors expert. The evaluator assumes the role of a naive end user, and will have had no previous exposure to the application. This evaluation session will predict which areas of the application may cause the end user difficulty.

Following the initial evaluation, the team at this stage may recommend that changes are made before the demonstrator is fully end-user tested.

End-user testing can only take place after planning for the tests has been completed. The evaluation team will decide on different scenarios which the end user will have to carry out to fully test the functionality of different strands of the application, e.g. 'During your flight you wish to make a telephone call to number 123456, using credit card 654321.'

Planning also includes recruitment of the end users. This is normally done through a recruitment agency. The user testing environment will, where possible, mimic the environment of the users of the service. Sessions of 8 to 10 end users are normally set up at any one time. Normally, a democratic representative sample of end users is used.

The human factors facilitator will ask users to perform a certain task. The users are asked to talk aloud, so that their thoughts and decisions can be recorded. This is followed by group discussions on the usability of the application in delivery of the task. Discussions can also be targeted at the naming of particular categories, or the implementation branding system.

The full session is recorded to audio tape and video tape. Following the user testing sessions, the evaluation team will study these, and write a document outlining the good and bad areas within the demonstrator, and, where appropriate, will make recommendations for change.

These changes will be discussed with the customer. If any of the changes are agreed, these are reflected back into the visual storyboards and, the concept demonstrator. Further user testing may be required, if significant changes have had to be made.

2.3.3.5   Design brief

The design brief is a written, high-level definition of the application to be designed. It takes into account all the requirements of the customer, the functional specification of the platform, and consistency issues raised through the style-guide. The visual storyboards and demonstrator support this document.

The design brief, on agreement, is signed by the customer, and is the absolute basis for the development of the application. The relevant technical experts use it to produce the detailed technical storyboards, data models, logical designs and test cases for verification, validation and testing. It is the brief for multimedia designers to produce the navigational content and product content for the application.

The design brief contains information on:

- application overview — detailing the navigation hierarchy of the application;

- application functions — detailing the full depth of navigation in the application, normally through one strand of the application, entering and exiting the application, getting help, error and information messaging, time outs;

- application contents — detailing fixed content, maintainable content and how regularly content will have to change;

- application branding — detailing the application styling and branding;

- usage context — detailing the user population;

- constraints — detailing constraints and restrictions on the application;

- deliverables — detailing deliverables;

- key dates and time-scales — detailing scheduled time-scales;

- usage statistics — detailing the usage statistics required by the service provider and information providers;

- acceptance criteria — detailing criteria for application acceptance, e.g. reviews and checkpoint meetings.

2.4   CONCLUSIONS

The service design process as embodied in Fig. 2.1, the 'service design pyramid', is fundamental to the successful design of a multimedia service and must be complete before the commencement of application development.

There must be sufficient documented requirements from potential information providers for each of the business strands within a service to enable the hardware and software platform to be selected.

The style-guide document should be substantially complete before starting application development and must document service values and give precise details of the branding system. The consequences of not doing so, could result in major rework of application designs with severe implications on time and budget.

3

GENERIC ASPECTS OF MULTIMEDIA PRESENTATION

M D Eyles

3.1   INTRODUCTION

The rapid development of multimedia computing has, to date, been driven by the impact of new hardware and software technology in fields such as storage, compression, networks and computing power. Although this trend is continuing apace, it is only recently that technology has matured to the extent that mass-market applications are becoming viable.

Traditionally, multimedia applications have been targeted at niche markets such as training, corporate presentation and desktop publishing. Technology solutions in these areas have been diverse, non-standardized and have failed to provide integrated and stable platforms for application development. However, ongoing technological change and standardization have moved the focus of multimedia applications towards the mass market and in doing so have changed the emphasis of development. Tomorrow's multimedia platforms will be consumer items aimed at the residential 'lounge' environment and will offer a large user base together with a stable and well-integrated platform for the deployment of new and innovative multimedia services. In particular, the emergence of the multimedia PC equipped with CD-ROM, sound card and video capability has high-lighted the potential consumer interest in multimedia products. Ongoing development of such technology is expected to include the convergence of PC, TV and communications services within general-purpose devices such as TV set-top boxes (although the extent and nature of this convergence is still hotly debated) (see Chapter 6).

With the emergence of mass-market multimedia systems, it is increasingly apparent that the value of a multimedia application will be perceived not as a function of the technology platform, but more as a function of the information available through the application. In this regard it is clear that content provision (in particular, aspects of production, and copyright), traditionally the domain of the entertainment industry, will be a major source of both costs and revenue. Hence, the production of content that is reusable across applications and services will be a major factor in the reduction of costs and in increasing profits.

Content production and reuse impacts on a number of key technology areas, such as multimedia hardware platforms, authoring systems, storage and distribution techniques and media-encoding formats. This chapter focuses on one particular area, the representation of complex multimedia content which may be reused across applications and services. It examines the reasons why a standardized approach for representing multimedia content is desirable, and discusses the concepts supported by such an approach with reference to the emerging ISO MHEG (Multimedia and Hypermedia Information Coding Expert Group) standards [1]. In addition, the chapter briefly illustrates the nature of future multimedia architectures based around reusable multimedia content.

3.2   THE NEED FOR STANDARDIZED MULTIMEDIA CODING

A simplified multimedia architecture is shown in Fig. 3.1. This architecture is typical of historical multimedia systems in which a vertical segmentation of functionality is present throughout the system. In such a system inter-application communication is either impossible or requires much effort to transcode between data formats, media encodings and command sets which are uniquely defined within each application context. Furthermore, in such a scenario, hardware environments are often widely incompatible making delivery of an application across multiple platforms costly.

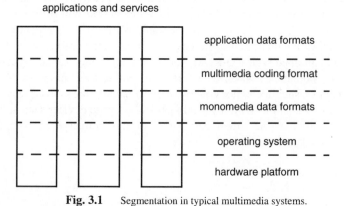

Fig. 3.1   Segmentation in typical multimedia systems.

The advent of standardized hardware platforms and operating systems, such as the multimedia PC, has improved interworking of applications by providing a stable and relatively mature hardware base around which applications may be developed. This has been the first stage in the rationalization of multimedia systems and a marked step towards inter-application operation.

The next step on the road to application interworking has been the development of standardized monomedia encodings. Typically such developments have been the role of *de jure* standards bodies such as ISO (International Standards Organization) and the ITU (International Telecommunications Union), and have resulted in technologies such as MPEG (Moving Picture Experts Group) video compression, JPEG (Joint Photographic Experts Group) image compression and other media encodings. Furthermore, many proprietary encodings have also been developed or are used for historical reasons. Typically, proprietary systems are aimed at specific hardware and software platforms (e.g. PC running Microsoft Windows™).

However, it is only relatively recently that the need for standardized multimedia coding schemes has been recognized. The advent of hardware capable of supporting and presenting complex multimedia content consisting of multiple media types (text, image, audio, video) has highlighted the requirement for reuse not only of monomedia data, but also the multimedia structures that surround such data. Indeed, it is clear that considerable effort, and hence value, is invested in the production of a multimedia presentation.

A successful multimedia coding scheme should support facilities applicable to a wide variety of applications and services and should result in widespread reuse of multimedia intellectual property. In this scenario, the simplified architecture is as illustrated in Fig. 3.2.

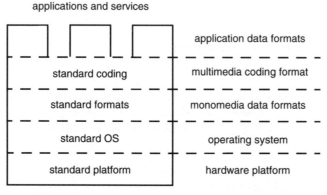

Fig. 3.2   Standard monomedia and multimedia data formats.

3.3   TECHNIQUES FOR MULTIMEDIA CODING

At present a wide variety of techniques are available for creating and representing multimedia content ranging from traditional DTP packages, 4GL programming languages and fast prototyping languages, through to complex middleware with associated multimedia scripting tools, databases, presentation and storage sub-systems. However, many of these systems blur the distinction between multimedia representation and the application, resulting in the production of multimedia content which is either fully encapsulated within a specific application or is not reusable across applications. Furthermore, multimedia content produced by these tools is typically coded in a proprietary format and hence is inextricably bound to particular software sub-systems and hardware platforms.

The problem of producing a non-proprietary multimedia content encoding, which is supportable across a wide variety of platforms, has recently been addressed within the ISO MHEG standards body. The aim of this body has been to produce a coding scheme for multimedia information which supports:

- application generic multimedia presentation concepts;

- fully self-defined complex multimedia presentations;

- multimedia content reuse;

- independence of application context;

- independence of hardware and software platform;

- independence of distribution mechanism;

- presentation on restricted resource hardware platforms;

- established monomedia coding formats;

- an open coding scheme.

The initial phase of this work is now complete and MHEG Part 1 (Coded Representation of Multimedia and Hypermedia Information Objects) is now an international standard. Further developments of the MHEG standard now include:

- MHEG5, 'Support for Base Level Interactive Systems', focusing on multimedia delivery to STBs;

- MHEG6, 'Support for Enhanced Interactive Applications', focusing on the provision of an interface between MHEG5 and a Java™ virtual machine.

3.4   THE MHEG MODEL

The MHEG model is a simple one — there should exist a standardized coding scheme for multimedia data structures which can be incorporated into diverse applications, services and platforms. The coding scheme should offer facilities to represent concepts which have been identified as being generic to a wide variety of applications and that are sufficiently flexible to support complex multimedia presentation. In addition, the coding scheme should be applicable to a wide range of platforms, storage and distribution techniques with a particular emphasis on presentation by restricted resource devices (e.g. TV set-top boxes) and distribution via telecommunications networks. The MHEG Part 1 Standard provides a public specification for the multimedia data structures developed using ASN.1 (Abstract Syntax Notation 1). The ASN.1 notation provides a specification of complex nested data structures which may be encoded to a bit stream according to standard rules. These bit streams may be interpreted by any general-purpose computer equipped with an ASN.1 decoder and hence act as an intermediate transfer format for multimedia objects between different internal coding schemes. The basic MHEG model is illustrated in Fig. 3.3.

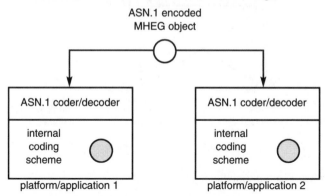

Fig. 3.3   Basic MHEG model.

Figure 3.3 shows that multimedia data structures held within platforms and applications may be encoded in a private or proprietary format applicable to the particular processing environment. Reuse and platform/application interworking is achieved by encoding internal multimedia formats to and from the standard MHEG format. It is essential therefore that the MHEG coding scheme is sufficiently flexible to support most application requirements for multimedia presentation.

It should be noted that the MHEG model does not assume any particular set of presentation, processing, storage or distribution techniques for the encoded multimedia data.

3.5   GENERIC CONCEPTS IN MULTIMEDIA CODING

The following sections detail some of the key concepts of multimedia presentation which should be supported by a generic multimedia coding scheme.

As mentioned previously, existing multimedia coding schemes are in general inextricably bound within application or platform contexts. They may offer a subset of these key concepts or entirely different approaches to multimedia presentation. However, it is believed that the concepts presented here are sufficient to support a wide range of typical multimedia applications.

3.5.1   Multimedia object life cycles

During its lifetime, a multimedia object may exist in a variety of states. These states and possible changes between them are illustrated in Fig. 3.4.

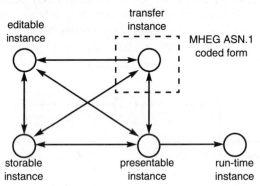

Fig. 3.4   Multimedia object life cycle.

The states of a multimedia object are fivefold.

- The editable instance, existing in a possibly private format within an authoring or editing application — the coding scheme employed should be amenable to fast revision, update and presentation and will be compatible with local programming tools. The MHEG coded form is not generally considered suitable for editing.

- The storable instance, existing in a possibly private format within a storage medium — the MHEG coded form is considered suitable for storage purposes and is independent of storage media.

- The transfer instance, existing between applications and platforms — the MHEG standard's primary function is to provide a standard encoding for transfer.

- The presentable instance, existing in a possibly private format within a presentation device (e.g. PC, set-top box) — the coding scheme employed should be amenable to fast presentation and will be compatible with the local programming environment and presentation devices. The MHEG coded form is not generally considered suitable for presentation.

- The run-time instance, existing in a possibly private format within a presentation device — the run-time instance represents an individual presentation to the user of the presentable instance. Many run-time instances may exist (simultaneously) as representations of a single presentable instance, sharing the monomedia and multimedia structures contained in the presentable instance. The coding format employed should be amenable to fast presentation and compatible with the local programming environment and presentation devices. The MHEG coded form is not generally considered suitable for presentation.

A multimedia coding scheme should address the migration of a multimedia object between the various states and in particular between the storage, transfer, presentable and run-time states within an application. Sophisticated referencing mechanisms are required to support the interaction between these states (see section 3.5.4).

The MHEG standard deals explicitly with the transfer, presentable and run-time states and adopts the model illustrated in Fig. 3.5.

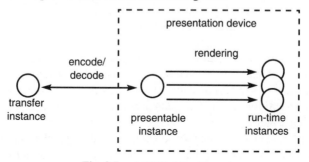

Fig. 3.5   MHEG object life cycle.

In this model, the transfer instance (encoded in MHEG form) is delivered to the presentation device. It is then the responsibility of the presentation device to decode the transfer instance to a suitable internal form, the presentable instance, ready for presentation. Many presentations of the presentable instance, the run-time instances, may then be made by rendering the monomedia data and multimedia structures expressed within the presentable instance.

The MHEG standard dictates how objects encoded according to the MHEG standard should be rendered by the presentation device.

3.5.2   Composition

A key function of a multimedia coding scheme is to describe the composition of individual media objects (text, image, graphics, video, etc) into complex multimedia structures and to control the relationships between the elements of these structures in terms of their presentation to and interaction with a user. This is illustrated in Fig. 3.6.

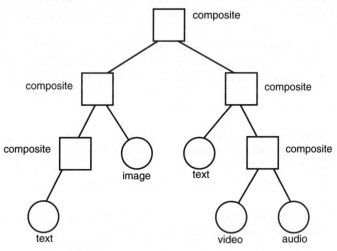

Fig. 3.6   Composition of multimedia structures.

Such structures may become arbitrarily complex with multiple levels of composition. Each element within the structure represents a reusable entity which (with or without its sub-components) may be transplanted to other structures. The basic requirement for multimedia content reuse is thus satisfied within the scope of the coding scheme.

The MHEG standard supports composition through the composite object structure, which is used to group sets of component objects and to express the rules and relationships governing their presentation and interaction.

In order to support composition and the meaningful presentation of multimedia content to a user, a generic multimedia coding scheme must offer a range of referencing, control, presentation and interaction facilities. These facilities may be used in the authoring stage to define multimedia content in terms of simpler monomedia and multimedia objects.

In particular, the following four facilities are required.

3.5.2.1   Support for multiple media types

The coding scheme should support the referencing and/or encapsulation of objects encoded according to existing monomedia encoding formats. These objects comprise the basic visible and audible elements of a multimedia presentation. Typically encoding formats are available for monomedia data such as video (MPEG, H.261, etc), image (JPEG, GIF, TIFF, BMP, etc) and audio (G.711, WAV, etc). However, consideration should also be given to support for complex media types such as data sets (spreadsheets, CAD files) and structured text (DTP, word processor output).

Two ways of integrating monomedia objects within a coded structure are illustrated in Fig. 3.7 — encapsulation and referencing. Encapsulation allows encoded media data to be carried within the monomedia object representing the data enabling the distribution of the monomedia object as a single entity. Referencing allows encoded media data to be stored or distributed separately from the monomedia object. This is of particular use where the referenced media data is large and may not be required for presentation or is to be streamed in real time from a data source (e.g. video camera).

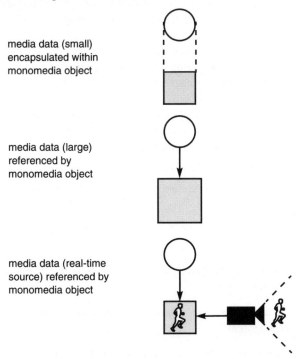

media data (small)
encapsulated within
monomedia object

media data (large)
referenced by
monomedia object

media data (real-time
source) referenced by
monomedia object

Fig. 3.7   Encapsulation and referencing of media data.

Additionally, the multimedia coding scheme should be able to uniquely identify each encoding to the presentation sub-system in order to ensure that the correct resources and devices are available for presentation of the media data. A standardized identifier for each data encoding is thus desirable to ensure interoperability and reuse between diverse platforms and applications.

The MHEG standard provides both referencing and encapsulation of monomedia data through standard referencing mechanisms. Data encoding identification may be achieved via a standard MHEG catalogue, private catalogues or appropriate externally defined catalogues.

3.5.2.2   Composition of multiple media

In order to provide true multimedia presentation (as opposed to merely presentation of multiple media types), the coding scheme must support facilities to relate individual media objects to each other in order to produce meaningful output. This is addressed through the composition function with relationships between media objects being expressed in the parent object (or script).

Various techniques exist for the specification of relationships between objects within a multimedia composition. These techniques are often tightly tied to the nature of the editing and presentation environment or to the requirements of specific applications. In particular, popular techniques that are employed range from basic spatial composition using drag-and-drop interfaces and basic temporal composition using timeline editing, through to complex multimedia scripting languages. Scripting languages typically offer near-natural language or visual specification of complex application logic and presentation control with tightly integrated hardware and software-presentation sub-systems.

It is possible to identify a number of key capabilities to support advanced multimedia composition:

- specification of spatial relationships formed by explicit links between objects;

- specification of temporal relationships formed by explicit links between objects;

- dynamic control of the spatial, temporal and audible parameters associated with objects in the system;

- run-time access to the state of, and parameters associated with, multimedia objects in the system;

- run-time capture of events associated with multimedia objects in the system;

- run-time capture of external events such as user interaction;

- support for logical combinations, conditions and tests of object state, object parameters, variables, internal and external events;

- support for flow control within scripts such as conditional execution, branching and loops.

The MHEG standard aims to provide a unified structure for the expression of relationships between multimedia objects. This structure, which is contained in a composite object, can be used to define dynamic spatial and temporal relationships between the components of the composite (and other objects) and to provide run-time access and operations on component state and component parameters. The MHEG Part 1 standard provides only minimal support for script-like flow control and complex logical and arithmetic operations. Such functionality is to be provided later in the MHEG scripting language.

The structure expressing inter-object relationships takes the form of a generalized hypermedia link and comprises a condition/action pair known as a link. Sets of links can be attached to objects (known as source objects). Each link expresses:

- a set of conditions in the form of logical and arithmetic tests on object state, object parameters and events within the system — conditions within the link are logically combined to form a single condition;

- a set of actions each targeted at specific multimedia objects within the system — the actions take the form of operations affecting object state, object parameters and causing new events to be generated.

Events (state changes, parameter changes, external events) within the multimedia system cause the evaluation of all currently active link conditions. A link condition which is satisfied causes the corresponding actions to take place on the targeted objects as illustrated in Fig. 3.8.

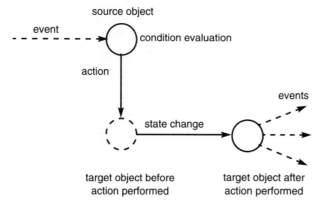

Fig. 3.8   Link effect between objects.

Typical uses for the link condition/action structure are detailed in the following sections on spatial and temporal composition in which links may be used to achieve visual integration and synchronization of media playback.

3.5.2.3   Spatial composition

Spatial composition (e.g. figure and caption, graphics overlay) is provided within an (X, Y) orthogonal co-ordinate system. Minimal facilities to support include:

- X, Y positioning (relative and absolute);

- scaling and clipping.

The MHEG standard provides an expressive set of mechanisms for defining the spatial relationships between media objects. These mechanisms may be used to produce complex visual effects and relationships between visual media objects, such as scrolling, clipping, scaling, grouping of objects and windowing effects. Relationships are defined using link constructs between pairs of objects. A simplified view of the features supported by the standard is given in Fig. 3.9 with links indicated by arrowed lines. The features illustrated comprise:

- X, Y positioning of a visible window on a media object relative to the parent object, sibling object or any other object within the presentation system;

- width, height specification of the visible window for clipping purposes;

- X, Y positioning of the media object relative to the visible window for clipping purposes;

- width, height specification of the media object for scaling purposes.

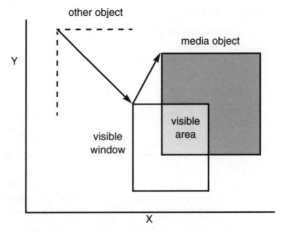

Fig. 3.9   Simplified view of spatial composition.

Spatial composition may occur at many levels and can lead to complex visual presentations. Additionally, the standard considers Z-axis positioning and transparency in order to achieve overlay effects between media objects.

All attributes of spatial composition may be controlled at run time with actions attached to links allowing dynamic visual effects to be realized. Typical action effects include altering the position and size of the visible window or media object.

3.5.2.4 Temporal composition (synchronization)

Temporal composition (e.g. audio commentary to slide-show, video subtitling) is provided along a single dimension co-ordinate system representing time. Facilities to be supported include:

- start/stop times for media playback,

- positioning/repetition of playback within media,

- speed variation.

The MHEG standard again provides an expressive set of mechanisms for defining the temporal relationships between media objects. These mechanisms may be used to control the relative speed, playback position and duration of media objects in addition to defining complex synchronization effects. Relationships are defined using link constructs between pairs of objects. A simplified set of examples of synchronization facilities supported by the standard are illustrated in Figs. 3.10 to 3.14 with links indicated by arrowed lines.

Temporal composition may occur at many levels and can lead to complex synchronization effects between many objects.

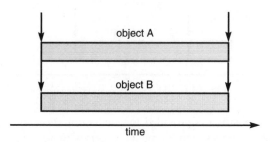

Fig. 3.10   Parallel playback of objects.

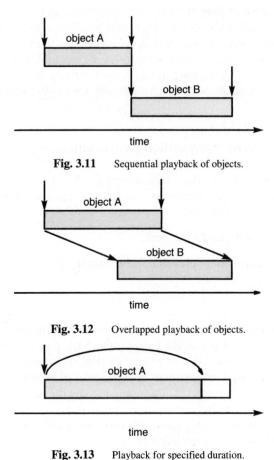

Fig. 3.11   Sequential playback of objects.

Fig. 3.12   Overlapped playback of objects.

Fig. 3.13   Playback for specified duration.

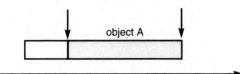

Fig. 3.14   Playback from specified position.

Temporal composition is provided for media objects with implicit duration (e.g. audiovisual sequences) and for media objects with no implicit duration (e.g. text, images) where the length of presentation is to be controlled.

All parameters of temporal composition may be controlled at run-time within an MHEG object allowing complex synchronization to be realized. A typical action effect is to vary the speed of playback, start times and stop times for media objects.

3.5.3   Referencing

The ability to accurately and uniquely reference objects within a multimedia system is a key requirement for the support of reuse and composition of multimedia data structures. The following facilities can be identified:

- referencing of media data on remote and local storage sub-systems — references must be sufficiently powerful and general to deal with a wide range of platforms, networks, filing systems and storage devices on which media data may be stored (note that this includes real-time media devices, e.g. cameras, microphones, in addition to traditional storage devices);

- referencing of storage instances stored on remote and local storage sub-systems — the ability to uniquely identify storage instances is essential for reuse;

- referencing of presentable instances within the presentation environment by the using application and by other objects — each presentable instance within the presentation environment must have a unique identifier to allow the application or other objects to request preparation, presentation or removal of the object and associated resources;

- referencing of run-time instances of multimedia objects within the presentation environment by the using application and by other objects — each run-time instance must be uniquely identified to allow the using application or other objects to control the presentation of that object;

- referencing of the components of multimedia composition — each component within a composite object should be addressable relative to the composite object; components should be addressable in their storage, presentable and run-time states.

The requirement for referencing of multimedia objects and the components of a multimedia composition across storage, presentable and run-time states together with media data is illustrated in Fig. 3.15.

Referencing of component objects relative to a parent (or ancestor) object is illustrated in Fig. 3.16.

The MHEG standard provides support for all the referencing techniques identified above and relies heavily upon them to achieve multimedia data structure reuse, composition and complex presentations.

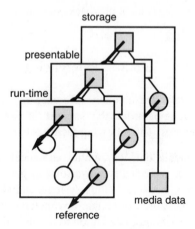

Fig. 3.15   Referencing in a multimedia system.

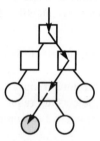

Fig. 3.16   Referencing a component object.

3.5.4   Interaction

Multimedia applications are typically highly interactive and the requirement to reuse multimedia data content extends across interaction with the user as well as presentation to the user. A number of facilities can be identified which enable the concepts of interaction to be integrated within a multimedia coding scheme:

- interaction objects which may be mapped to typical modes of user input (text input, menus, buttons, pointers, check boxes, scroll bars, gauges);

- integration of interaction objects within the composition process in terms of spatial and temporal relationships with other objects;

- interrogation of the state of interaction objects — links representing logical relationships with other objects may be activated dependent upon the state of an interaction object;

- interrogation of the state of interaction objects by the using application in order to pass user input to the application for application-specific processing.

The ability to support interaction objects within the composition structure allows the definition of reusable user interface components which may be applicable to a wide range of applications.

The MHEG standard supports the composition of interaction objects within multimedia data structures and enables the definition of reusable user interface components which are independent of application. Composite objects may contain links which perform interrogation and logical testing of the state of interaction objects and cause appropriate actions to occur.

3.5.5   Example multimedia object

This section illustrates the generic multimedia concepts outlined above through the construction of an example multimedia object. The example is intended to clarify the concepts discussed but is not intended to represent a particular multimedia coding scheme (e.g. MHEG).

The example multimedia object will represent a commonly used sub-component of a multimedia presentation — a subtitled audiovisual clip with VCR-like user control. This example provides an opportunity to illustrate:

- referencing of monomedia data;

- encapsulation of monomedia data;

- referencing of monomedia components;

- composition of a multimedia structure;

- spatial relationships;

- temporal relationships;

- interaction facilities;

- progression through the life cycle.

Construction and presentation of the example multimedia object is illustrated in the following five phases.

Phase 1 — Fig. 3.17 illustrates the referencing and encapsulation of monomedia data within simple multimedia component objects. In this example, large files, such as video data and image data, are referenced from the component objects while small files, such as text subtitles are encapsulated within the component objects.

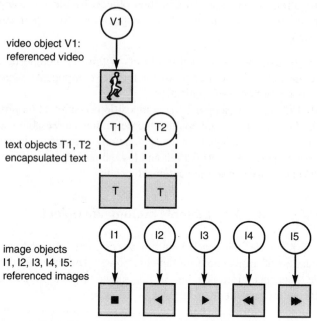

Fig. 3.17   Referencing and encapsulation of media data.

Phase 2 — Fig. 3.18 illustrates the composition of text and video components within a composite object representing the subtitled video and image components within a composite object representing the VCR controls. In this case, the monomedia components are referenced by the containing composite objects.

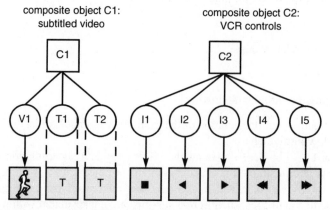

Fig. 3.18   Composition of component objects.

Phase 3 — Fig. 3.19 illustrates the composition of the subtitled video object and the VCR control object within an overall composite object for presentation purposes. It should be noted that as the subtitled video object is referenced by the higher level composite, it would be possible to exchange the subtitled video with another of similar structure. This allows the VCR control object to be used to control any number of subtitled video objects. The MHEG standard provides facilities to plug and unplug sub-structures from the composition to achieve such effects.

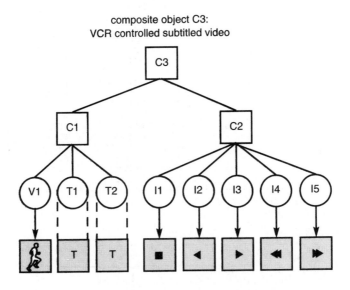

Fig. 3.19   Higher levels of composition.

Phase 4 — Figs. 3.20a to 3.20c illustrate the construction of a simplified control script to be attached to the composite objects within the structure. This structure takes the form of a number of condition/action pairs which are used to monitor events in the system and take action as required.

Fig. 3.20a shows actions to be taken when a using application requests the preparation of the composite object structure for presentation. The specified controls cause the preparation of sub-components within the hierarchy and the setting of initial position, size, temporal position and speed values for each of the monomedia components. In addition, the relative positions of the composites C1 and C2 with respect to the composite C3 are specified. The resulting spatial relationships are as shown.

Source	Condition	Target	Action	
C3	Prepared	C1	Prepare	
		C2	Prepare	
		C1	Set Position	100, 100
		C2	Set Position	100, 420
C1	Prepared	C1.V1	Set Position	0, 0
		C1.V1	Set Size	400, 300
		C1.V1	Set Time	0
		C1.V1	Set Speed	0
		C1.T1	Set Position	200, 250
		C1.T1	Set Size	200, 50
		C1.T2	Set Position	200, 250
		C1.T2	Set Size	200, 50
C2	Prepared	C2.I1	Set Position	0, 0
		C2.I1	Set Size	80, 80
		C2.I2	Set Position	80, 0
		C2.I2	Set Size	80, 80
		C2.I3	Set Position	160, 0
		C2.I3	Set Size	80, 80
		C2.I4	Set Position	240, 0
		C2.I4	Set Size	80, 80
		C2.I5	Set Position	320, 0
		C2.I5	Set Size	80, 80

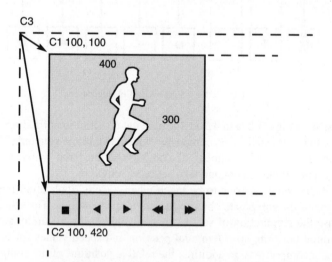

Fig. 3.20a   Specification of initial spatial relationships.

Fig. 3.20b shows actions to be taken when a using application requests that the composite object structure be presented. The specified controls cause the start of sub-components within the hierarchy, resulting in presentation of the video and VCR controls.

Source	Condition	Target	Action
C3	Started	C1	Start
		C2	Start
C1	Started	C1.V1	Start
C2	Started	C2.I1	Start
		C2.I2	Start
		C2.I3	Start
		C2.I4	Start
		C2.I5	Start

Fig. 3.20b   Presentation of composite object.

Source	Condition	Target	Action	
C1.V1	Time=5	C1.T1	Start	
	Time=15	C1.T1	Stop	
	Time=25	C1.T2	Start	
	Time=30	C1.T2	Stop	
	Stopped	C1.T1	Stop	
		C1.T2	Stop	
C2.I1	Selected	C1.V1	Set Speed	0
		C1.V1	Stop	
C2.I2	Selected	C1.V1	Set Speed	−100
		C1.V1	Start	
C2.I3	Selected	C1.V1	Set Speed	100
		C1.V1	Start	
C2.I4	Selected	C1.V1	Set Speed	−500
		C1.V1	Start	
C2.I5	Selected	C1.V1	Set Speed	500
		C1.V1	Start	

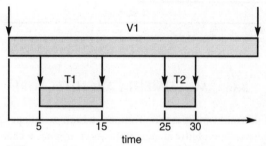

Fig. 3.20c   Specification of temporal synchronization and control

Fig. 3.20c shows actions to be taken during the presentation of the composite object structure. In particular, actions are present to control the playback of the text subtitles at the correct points in the video with the timeline synchronization as shown. Additionally, actions are provided to control the speed of video playback via the VCR control buttons which may be selected by the user.

Phase 5 — Fig. 3.21 illustrates the production of two separate run-time instances (e.g. at the request of an application) of the same composite object structure. It should be noted that as every spatial and temporal positioning specified within the composite object structures is relative, then only positioning of the outermost object C3 is required.

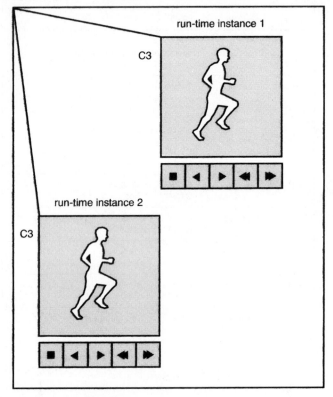

Fig. 3.21   Creation of run-time instances.

3.6   MULTIMEDIA ARCHITECTURE

The advent of reusable multimedia content, which is independent of storage, distribution, presentation platform and application, implies a change in the nature of multimedia architectures. As previously indicated, existing architectures blur the distinction between content and application and rely on 'stove-pipe' hardware and software systems. These systems are highly integrated and dependencies exist between all layers from application to hardware. Newer architectures are beginning to rely on the development of standardized interfaces between key components of the architecture. Standardized multimedia information content allows this process to move a step further.

Some of the key components to be found in a multimedia architecture developed around reusable multimedia content are listed below. A simplified architecture together with potential standardized interfaces between the components is illustrated in Fig. 3.22.

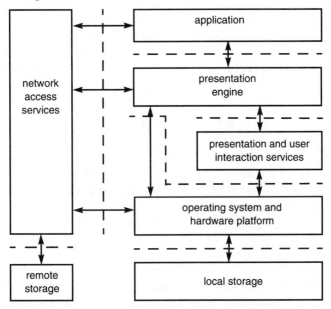

Fig. 3.22   Example multimedia object with interaction.

The architecture comprises:

- the application, providing context-specific processing and interfacing to the underlying presentation engine — the application is able to control the presentation and interrogation of multimedia objects through this interface;

- the presentation engine, providing processing of multimedia objects encoded according to the standard coding scheme — the presentation engine has access to remote and local storage services through the network and local hardware platforms respectively, and is responsible for the processing of the composition rules expressed within the multimedia objects and for issuing requests to the underlying presentation services;

- the presentation and user interaction services, providing media presentation and user interface facilities through the underlying operating systems and hardware platform;

- local and remote storage, providing storage for multimedia content and media data;

- network services, providing access to remote storage and advanced facilities such as object location services and quality of service negotiation.

A key requirement is the definition of standard interfaces between the application and presentation engine and between the presentation engine and presentation services. Definition of these interfaces will allow the development of 'black-box' presentation engines that are able to process multimedia content encoded according to the standard coding scheme. Development of such engines for a wide variety of platforms will enable inter-application and inter-platform multimedia communications on a wide scale.

3.7   CONCLUSIONS

This chapter has briefly presented some of the basic concepts associated with complex multimedia presentation and the requirements for a generic and standardized multimedia coding scheme. In addition, architectural considerations for future multimedia systems have been presented. Particular reference has been made to the emerging ISO MHEG (Multimedia and Hypermedia Information Coding Expert Group) standard as an example of how such concepts and architectures may be provided.

It is anticipated that standardized coding of multimedia content will become a key technology with the emergence of mass-market multimedia applications and services. In particular, the costs associated with multimedia content production are expected to highlight the need for a widespread coding scheme allowing content reuse.

It is clear that content providers, service providers, application developers, network providers, software and hardware manufacturers all need to consider these issues if widespread uptake of multimedia applications and services is to be achieved.

REFERENCES

1. ISO/IEC IS 13522-1: 'Information Technology — Coding of Multimedia and Hypermedia Information — Part 1: MHEG Objects Representation Base Notation (ASN.1).

    ISO/IEC DIS 13522-3 Information Technology — Coding of Multimedia and Hypermedia Information — Part 3: MHEG Script Interchange Representation.

    ISO/IEC IS 13522-4 Information Technology — Coding of Multimedia and Hypermedia Information — Part 4: MHEG Registration Procedure.

    ISO/IEC DIS 13522-5 Information Technology — Coding of Multimedia and Hypermedia Information — Part 5: Support for Base Level Interactive Applications.

    ISO/IEC WD 13522-6 Information Technology — Coding of Multimedia and Hypermedia Information — Part 6: Support for Enhanced Interactive Applications.

4

SERVICES, TECHNOLOGY AND STANDARDS FOR BROADCAST-RELATED MULTIMEDIA

W Dobbie

4.1   INTRODUCTION

Broadcast television is perhaps the most widely accepted form of multimedia communication today. Since its introduction it has had an impact on social, economic and political development world-wide at least comparable with that of the telephone. The more recent introduction of practical video recorder/players has added the capability to view broadcast services at a more convenient time and to view non-broadcast material which is purchased or rented from a wide range of outlets on impulse. This complementary facility has been widely adopted by consumers, as has the remote control. Teletext was not, however, taken up quite so enthusiastically and Prestel was even less well received which shows that it is important to ensure that consumers will find a new offering attractive before committing to it.

Broadcast technology is developing rapidly and the change from analogue to digital TV delivery is almost upon us. The main advantage for conventional broadcasters will be increased capacity and flexibility and this will be used to increase the range of programming to cover a wider range of information, entertainment and transactional services and to offer increased control through 'staggercasting' which is a system where material is transmitted many times with relatively short time delays between each broadcast. There are clear plans to use staggercasting for popular movies and this is referred to as near video-on-demand (N-VoD), but the same techniques can also be applied to news and other material.

It remains to be seen, however, how much extra money consumers will be prepared to pay for these new services and for the equipment to receive and decode them. There is therefore considerable interest in adding an extra dimension to the new services by enabling network interactions to be associated with broadcast services in a practical manner. This would combine the functionality of the existing telephone network with that of the TV set to allow audio, text and still-picture information to be provided on demand.

This could provide a practical and attractive stepping stone towards other new network services such as videophone and video-on-demand by building up consumer and information-provider awareness of the ways in which a TV display and a remote control can be used for personalized services. It will, however, be necessary to learn from the experiences with Prestel and ensure that the services are appropriate for a TV environment rather than a personal computer environment.

This paper mainly concentrates on the specifications which are being developed by the European Digital Video Broadcasting (DVB) project to enable delivery of new digital broadcast TV services (including interactivity as an option) in a uniform manner over all practical delivery platforms to low-cost set-top boxes. The benefits from the resulting mass market will enable manufacturers to provide a range of interoperable equipment (at least for satellite and digital terrestrial TV) for sale in the high street.

Sections 4.2 and 4.3 set the scene by summarizing the different delivery options in terms of current analogue systems and the progress on development of digital TV standards world-wide. Section 4.4 presents the main achievements of the DVB project in greater detail and section 4.5 discusses the requirements and the system implications for addition of PSTN- or ISDN-based interactivity.

4.2   ANALOGUE TV BROADCASTING

Broadcast TV is normally formatted using PAL or Secam in 625 line/50 Hz countries or NTSC in 525 line/60 Hz countries. In each case there are many regional variations which require the provision of different equipment. The lack of a single analogue standard also makes it necessary to convert material in many cases, which causes a loss of picture quality.

Figure 4.1 illustrates the three delivery options used to provide analogue broadcast TV to the majority of viewers in the UK, and they are described below.

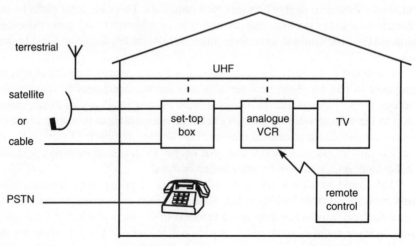

Fig. 4.1   Analogue delivery.

- Terrestrial UHF — signals are normally transmitted in a frequency division multiplex (FDM) using vestigial sideband amplitude modulation (VSB/AM) and received using an antenna such as a Yagi array. The UHF receivers are integrated into TV sets. Manufacturers have tried for many years to convince consumers to use the higher quality baseband interface which is now fitted on most new TV sets, satellite set-top boxes and VCRs (referred to as SCART), but they have found in many countries including the UK that they had to include a modulator to convert the baseband signals to UHF, presumably because the facilities provided by these appliances are seen as complementary to the basic terrestrial broadcast services.

- Satellite — signals are normally transmitted in FDM using frequency modulation (FM) with one TV signal occupying a complete satellite transponder (effectively an amplifier and frequency converter) and received using a parabolic dish.

  Significant effort was invested (primarily by manufacturers) in the development of an alternative to PAL/Secam for satellite broadcasting (D/D2-MAC within Europe). This system used multiplexed analogue components (MAC) to improve the picture quality for conventional TV services and to allow for the future introduction of high-definition TV (HDTV) in a compatible manner. It was not adopted because consumers were at that time more interested in paying for the boxes and subscriptions to receive extra TV channels than for increased quality. There was also by then the prospect of digital TV which will offer further increased capacity and flexibility and allow for compatible HDTV.

Initially, many of the English language satellite TV channels were funded by advertising. Although there are still a few of these, the vast majority are now funded by subscriptions. Over 3.5 million consumers pay for access to the BSkyB services and most of these receive the signals directly from satellite. Conditional access control, based on over-air authorization of SMART cards in set-top boxes, is used to ensure that consumers pay.

• Cable — signals are normally transmitted in FDM using VSB/AM via a coaxial cable. In most cases a large dish is used to receive satellite services which are then passed on to consumers. In many cases this means that the cable TV services are more expensive than direct satellite reception; however, the cable companies can use different packages of programmes and charging arrangements. Some consumers have found this flexibility, plus the avoidance of the need for a dish, attractive. Nevertheless, the take-up of cable was relatively slow until the UK regulations were changed to allow cable TV franchise holders also to offer telephony.

Considering the rest of Europe, satellite master antenna (SMATV) networks are also used (particularly in Spain and Italy) to distribute satellite and terrestrial TV signals from a shared dish or dishes throughout blocks of flats using coaxial cables. The signals are normally converted to VSB/AM for transmission, but in some cases the FM satellite format is retained to allow use of standard satellite set-top boxes.

In the USA and Ireland, microwave video distribution (MVDS) is used with VSB/AM as an alternative to cable, once again primarily for distribution of satellite TV services. In this case the signals are transmitted from masts in a similar way to conventional terrestrial TV but at around 2.5 GHz. Modified cable TV set-top boxes are used. BT has also investigated the use of MVDS at 40 GHz with the FM satellite format and a major customer trial was performed at Saxmundham during 1988 [1]. MVDS was also proposed for use in the West Kent cable TV franchise area.

Figure 4.1 also shows a telephone which represents the most practical means of interacting with the broadcast service. Telephones and the PSTN are increasingly being used to respond to talent and gameshows and to order goods advertised via the TV channel or associated teletext services. In most cases the interaction is via an operator, but there are also cases where the telephone key pad is used.

A number of cable TV companies in the USA, Canada and Europe are providing interactive services including pay-per-view, teleshopping and games using dedicated low-bit-rate data modems to respond to a control computer over the coaxial cables.

4.3   EVOLUTION OF DIGITAL TV STANDARDS

In the late eighties, after many years of research into digital TV coding, developments in motion adaptive discrete cosine transform processing and VLSI implementation first made it possible to achieve a picture quality acceptable to most viewers at relatively low bit rates and potential implementation cost. This advance, plus the parallel developments in low-bit-rate audio coding, provides increased flexibility in terms of the bandwidth required for services plus the benefits of digital coding for transmission and volume production for the first time.

The new digital technology was initially exploited by companies involved with satellite and cable delivery in the USA including General Instruments and Scientific Atlanta. They rapidly developed and implemented proprietary hardware and some of this was put into service. During 1990 General Instruments also proposed the use of digital coding for terrestrial transmission of HDTV within the USA and this was enthusiastically adopted by manufacturers who formed the 'Grand Alliance' to jointly develop a standard.

The development of the early systems was progressed in parallel with the work of the Moving Picture Experts Group (MPEG), often by the same engineers. Many systems were therefore based on the MPEG-1 specification which, although primarily intended for video disk encoding, can also be used at higher rates. RCA (a US subsidiary of Thomson Consumer Electronics) extended this ISO standard and implemented low-cost decoders for the Direct TV satellite broadcasting system. This 150-channel service became operational early in 1994 and has been extremely well received by consumers throughout the USA, even in areas served by cable TV networks.

During 1994 the MPEG-2 system for broadcast TV was also finalized [2—4]. Most companies have now updated their designs to include the enhanced coding and the multiplex formats which it provides and it seems extremely likely that it will steadily become the world standard for digital TV broadcasting. This has allowed the major silicon companies to invest in VLSI chips with confidence and there are already chip sets available. Most companies also have plans for steadily increasing levels of integration which will result in considerable savings on set-top boxes (see Chapter 6). Services such as Direct TV, which committed to silicon before the finalization of MPEG-2, will remain incompatible for some time, but some of these companies are now planning to change when further generations of equipment are introduced.

Although the MPEG-2 world standard for coding and multiplexing provides an excellent basis for the introduction of digital TV broadcasting there are, however, many other aspects which have to be addressed and, if possible, standardized. Figure 4.2 illustrates the most likely scenario for the introduction of digital TV broadcasting.

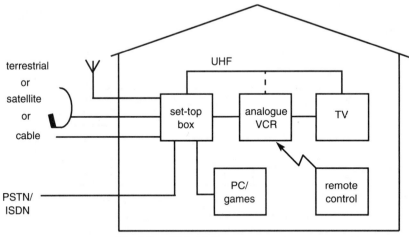

Fig. 4.2   Digital delivery — near-term.

The same three delivery options are likely to remain dominant within the UK, and in the near-term it is likely that dedicated set-top boxes that include a network-specific receiver/decoder will be provided. Although it would be possible to include more than one receiver/decoder in a box this is not likely to be the norm.

In the short-term, it is likely that the output interfaces will remain analogue because consumers will wish to use existing VCRs and TV sets. This could be achieved by PAL encoding and then modulating in VSB/AM on to a UHF carrier, as for current analogue set-top boxes, or by connecting the baseband signals via SCART cables which would preserve the full quality.

The boxes will in many cases include a digital interface to allow data including software and games to be downloaded to local stores in personal computers (PCs) or games machines. This also allows multimedia services, including Internet and other specialist news, to be downloaded in a cost-effective manner, possibly overnight. This would, however, require wiring between rooms and the effort and cost involved may discourage adoption. The boxes will also, in most cases, have a PSTN (or possibly ISDN) modem incorporated to allow for a range of interactive services associated with broadcasting, as described in section 4.5. This could in some cases also be used in conjunction with the PC, subject to the same wiring requirements.

Figure 4.3 shows how the home entertainment environment could develop. As MPEG-2-based services become established, manufacturers will increasingly incorporate the receiver/decoders for the most common delivery systems into new TV sets, possibly as plug-in network interface units, purchased and added as

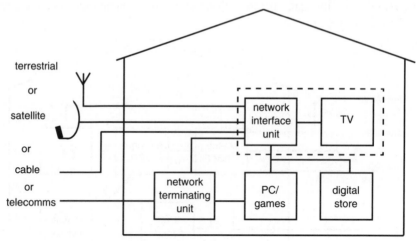

Fig. 4.3   Digital delivery — longer term.

required. This could help to sell sets for the wide aspect ratio HDTV services which are likely to be steadily introduced. The savings in avoiding one or more set-top boxes would help to offset the extra cost of the wide aspect ratio TV when a new set is required.

These digital systems represent an efficient way of broadcasting popular material. Combining these with the PSTN (or even ISDN) will allow a degree of interactivity and personalization to be added to broadcasting and its use is expected to build up over time. Ultimately, switched broadband networks could be used to provide broadcast and on-demand services chosen by the individual as an alternative or supplement to other services. Such services are already in the early stages of commercial trials (see Chapter 6).

The other major change is likely to be the implementation and acceptance of a digital alternative to the analogue VCR. Digital VCRs have been specified by a world-wide consensus-forming body termed the DVC Conference. This specification is intended to allow for analogue inputs and outputs and MPEG coding is not used. It can, however, also be used to record MPEG-encoded signals if a suitable digital interface is provided. The bit rate recorded on the tape is relatively high (25 Mbit/s for conventional TV), partly because the coding is less efficient than MPEG but also because it is necessary to allow a significant overhead to permit fast forward and other VCR functions to be provided.

A specification has also now been produced for digital VHS recording and playback of MPEG or other data streams and it seems likely that this will allow lower cost implementation for similar functionality. There is also the prospect of recordable disks and even solid-state memories with enough capacity to store a movie. It is very difficult to predict which technology will eventually replace the analogue VCR but it is clear that an interface for a local digital storage capability will be needed.

Two major consensus forming bodies are developing specifications for digital TV broadcasting and the physical and non-physical interfaces required to evolve to the situation illustrated in Fig. 4.3.

4.3.1   European Digital Video Broadcasting (DVB) project

The European DVB project was officially inaugurated in September 1993 [5 ,6]. It grew from the 'European Launching Group for Digital Video Broadcasting' which had already performed a considerable amount of work, particularly on digital terrestrial TV. The project currently consists of a voluntary group of around 220 organizations, including BT, who have signed a 'Memorandum of Understanding' describing the goals of the project. These are to enable the development of standards for digital video broadcasting covering all delivery options and the early introduction of services while avoiding the pitfalls experienced with analogue TV.

The project is funded by the members and has developed its own objectives, policy and rules of procedure, acknowledging that the new digital broadcast and electronic multimedia environment requires market-led approaches to technical development. The members represent four main constituencies, namely consumer electronics manufacturers (including many US and Japanese companies with European subsidiaries), broadcasters and programme providers, network and satellite operators, and regulatory bodies including the European Commission. There is a Steering Board which controls three Commercial Modules dealing with Cable and Satellite, Terrestrial and Interactive services, an *ad hoc* group which has developed a policy for conditional access control and scrambling, and the Technical Module which takes 'user requirements' from the commercial modules and generates the specifications to implement them.

The DVB project is not a standards-setting body but it has formed a close liaison with the European standards organisations (ETSI and CENELEC) based on a formal co-operation contract. The main achievements of the DVB project to date are described in section 4.4. The decision to add interactivity to the DVB specification was finally made by the Steering Board (following requests from BT and others) during March 1995 [7]. The ongoing and planned work on this topic is described in section 4.5.

4.3.2   Digital Audio-Visual Council (DAVIC)

DAVIC is a world-wide consensus-forming body which was formed during 1994 to extend and utilize the work of MPEG. It has around 160 member organizations including large numbers from the USA and Japan where it has a very high profile. The goal is to define specifications for open interfaces and protocols for all audiovisual applications and services based on digital coding and delivery to maximize interoperability across countries and applications/services.

Once again this project is funded by the members who mainly represent manufacturers (mostly computer and cable TV, but also consumer electronics) and network operators, including telecommunications companies such as BT and a few satellite operators. There are relatively few broadcasters or programme providers and, recognizing this, DAVIC is keen to attract extra membership from this community.

DAVIC has a Board of Directors which is advised by a Management Committee and a Strategic Planning Advisory Committee. The Management Committee controls the work of six Technical Committees dealing with Applications, Sub-systems, Information Representations, Physical, Systems Integration, and Security. DAVIC is similar to DVB in that it is not a formal standards organization. Its aim is to adopt relevant existing standards where possible, and, when it produces its own specifications, to submit these to the standards bodies for approval. At the time of writing DAVIC had applied to ISO and the IEC for a formal liaison arrangement similar to that between DVB and ETSI/CENELEC.

The main focus of the first specification (DAVIC 1.0), which was finalized in December 1995, was on video-on-demand. The specification differentiates between:

- low-layer protocols and physical interfaces,

- mid-layer protocols,

- high-layer end-to-end protocols.

The physical interfaces include specifications for satellite, asymmetric digital subscriber loop (ADSL), fibre to the kerb, hybrid fibre/coax and ATM delivery to the set-top box which would allow for in-home distribution. An interface is also specified between a network interface unit and a set-top box.

4.4   DIGITAL VIDEO BROADCASTING STANDARDS

The DVB specifications include elements which are applicable in all cases and others covering modulation and coding for the different delivery systems.

4.4.1   Common elements

The aim of these specifications is to maximize commonality so that manufacturers can reuse as many elements as possible across the complete range of delivery systems.

4.4.1.1 MPEG architecture

The MPEG-2 standard [2-4] includes a family of video coding standards [6, 8]. These are separated into levels which define the number of pixels and profiles which specify the processing used. DVB has selected the Main Level and Main Profile option which has 720×576 pixels with B-frame coding because it most closely matches the anticipated service requirements. It has also selected a subset from the remaining toolkit and this is described in a guidelines document (one of a set of DVB blue books) with which manufacturers will voluntarily comply in their development and implementation of set-top boxes and encoders. The remaining flexibility in terms of the coding and resolution still allows a number of options to be implemented ranging from limited-definition TV (LDTV) at bit rates as low as 1.5 Mbit/s to enhanced-definition TV (EDTV) with wide-aspect ratio at up to 15 Mbit/s. Audio-coding is according to the MPEG layer 2 (MUSICAM) specification.

4.4.1.2 Service information

DVB has defined a set of service information (SI) tables which are transmitted with broadcast services to assist with the implementation of user interfaces and this specification has now been standardized by ETSI and adopted by DAVIC [9]. The SI tables contain information about each specific programme, packages of programmes from a specific broadcaster, the current event within the programme and other information. This will enable a viewer to navigate through the many channels which will become available from different sources, not only on a channel-by-channel basis but also by selecting categories of services. Information will also be supplied to allow parents to decide on suitability of programming for children and for electronic programme guides and many other desirable new features.

4.4.1.3 Teletext

A specification for carriage of conventional teletext has been standardized by ETSI [10]. In practice, many broadcasters are likely to introduce new graphics-based equivalents.

4.4.1.4 Subtitling

A specification for subtitling including the capability to deliver station logos and bitmaps for other graphics applications is currently being finalized and it will be passed to ETSI for standardization in the near future.

4.4.1.5   Conditional access

Conditional access (CA) control is essential for broadcast subscription TV as it ensures that consumers pay for the services. In most cases, consumers are supplied with a SMART card when they subscribe to the services and this is then plugged into a CA module inside the set-top box via a hole in the case and authorized via the broadcast channel. The CA module allows for the secure distribution of decryption keys which can be used to descramble services. This is a commercially sensitive subject and DVB has not attempted to define a standard CA system. It has, however, defined a common scrambling algorithm and fully specified an interface allowing boxes to accommodate more than one PCMCIA-mounted CA module [11]. This specification is currently being considered for standardization by CENELEC. It has also been submitted to DAVIC.

4.4.1.6   Physical interfaces

DVB has defined a set of physical interfaces covering inputs from all of the different network options and outputs to the TV set and to other peripherals and this has been submitted to CENELEC for standardization [12]. The interfaces are optional, but, if an interface is used, then it should comply with the DVB requirements to ensure commonality. An RS232 data port has been defined for communication with a PC or other device.

A integral PSTN modem is specified to allow for the interactive services described in section 4.5. One or more of the following modes should be supported — automode selection ITU V.21, V.23 (1200/75), V.22 or V.22 bis transmission protocols. It is also recommended that V.32, V.32 bis and V.34 support should be included and that designs should not preclude the addition of future enhancements. A connector allowing for an external PSTN or cable TV modem is specified. Control is via the Hayes AT command set.

The interfaces required for digital video cassettes (DVCs) have also been investigated, as discussed in section 4.3. Initially, a number of serial data interfaces were considered, but, following reports of support from the DVC Conference for the IEEE P1394 specification, it was decided to adopt this. Work was then carried out to define how a DVB MPEG transport stream should be carried over this interface in the most efficient manner. This work will be extended to allow for the other digital storage systems discussed in section 4.3.

Although the IEEE P1394 specification is suitable for connections between clusters of equipment close to each other it requires repeaters every 4.6 m which makes it impractical for home distribution applications including connection of telecommunications services from a network terminating unit to a network interface unit, as shown in Fig 4.3. DVB has therefore also investigated the requirements for a longer reach data interface. Although a number of suitable candidates were identified, it was decided to define a network-independent interface allowing customers to connect a network-interface module if and when required. A set of commercial requirements have therefore been produced and work to specify the interface is to start soon.

4.4.2   Delivery systems

These specifications define the modulation and coding for all of the delivery options considered by DVB so far. The Reed Solomon outer forward error correction (FEC) coding, which is used to combat the effects of impulsive noise, is the same for all delivery options. A common concatenated inner FEC code is also used for terrestrial and satellite delivery. This eases the design of the VLSI chips and reduces the complexity of transconversion between the satellite and cable formats. In all cases a single MPEG transport stream is used to carry a flexible mixture of video, audio and data channels in a time division multiplex (TDM). This means that the interface between the receiver/decoder (or network interface unit) and the MPEG decoder/demultiplexer can be the same for all delivery options.

4.4.2.1   Terrestrial

Although digital terrestrial TV (DTT) was the first delivery system considered by DVB, the specification required significantly more effort than the other delivery media. The reason is that there is very little spectrum available for new digital services in many countries and it is therefore necessary to allow for a range of applications and introduction scenarios. The BBC has, however, now decided to proceed with trials of DTT which should lead to services during 1998. The Government has now also issued a White Paper with proposals for DTT introduction, using allocations which are not used for analogue services to avoid unacceptable interference. The digital signals cause less interference and are affected less by interference. It is predicted that 24 new TV channels could be introduced over most of the UK. This assumes that each DTT carrier is used for an MPEG transport stream at around 24 Mbit/s shared between four TV channels in TDM. This has given an increased impetus to the work and a specification has been passed to ETSI.

The current specification is based on coded orthogonal frequency division multiplexing (COFDM) with associated convolutional inner FEC coding. Three modulation options are available and the inner FEC coding can operate at a range of rates. The symbol rate available for the MPEG transport stream can also be varied. This makes it possible to trade off capacity against performance in a flexible manner (from around 6 to 32 Mbit/s depending on carrier-to-noise requirements), allowing the same system to be used for different coverage and interference scenarios.

The COFDM process requires significant complexity, and this means that the receiver/decoders will be more expensive compared to those for satellite and cable. The complexity is further increased because the specification allows for the guard interval options to provide for the use of single-frequency networks in other European countries. Nevertheless, the advantage of reception via existing (and well-accepted) Yagi arrays and the attraction of an enhanced 'free-to-air' service (licence and advertising funded), plus new subscription services (probably using the wide-aspect ratio format from the start) are likely to ensure the volumes required to rapidly reduce the costs.

4.4.2.2   Satellite

The DVB satellite TV modulation and coding specification has now been standardized by ETSI and adopted by DAVIC [13]. The system includes flexibility in terms of the inner FEC coding rate and the symbol rate to allow for a range of satellite transponder bandwidths (26 MHz to 72 MHz) and powers (49 dBW to 61 dBW). QPSK modulation is used for transmission [14].

A number of satellite operators are planning to provide capacity for direct-to-home services and these will mainly have bandwidths of around 33 MHz and power of around 52 dBW, allowing around 39 Mbit/s to be delivered via a 60 cm dish with an acceptable margin. A 45 cm squarial antenna could be used if a lower margin was accepted or if the data rate was reduced. It is worth noting that with a digital system the picture quality normally remains constant until the carrier level becomes too low, for instance during a rain fade, when it can fail completely. It is therefore necessary to use a higher margin than for analogue satellite TV which degrades more gracefully.

BSkyB has transmitted signals via the Astra 1D satellite for demonstration purposes and to allow set-top box manufacturers to ensure that their designs are compliant. Other similar test transmissions are likely and services are likely to start soon. BSkyB is reported to be planning a 200-channel English-language service via an Astra satellite at a new orbital location. In fact, services have already started in France, South Africa and the Far East.

4.4.2.3   Cable

The DVB cable TV modulation and coding specification has also now been standardized by ETSI and adopted by DAVIC [15]. It was completed after the satellite specification in order to ensure maximum commonality for simple transmultiplexing. The signals are transmitted using quadrature amplitude modulation (QAM). The normal mode is likely to be 64 QAM, but 16 QAM and

32 QAM are also specified and these may be used for SMATV, as described below. Higher rate systems are also defined, but their use will depend on the capacity of the cable network to deal with the reduced data eye height (and the fact that DAVIC has defined a different system for these rates).

It is possible to vary the symbol rate (the maximum rate will be around 39 Mbit/s for 64 QAM) as for satellite, and this makes it possible to assemble multiplexes including TV channels from many sources with a different conditional access control system to that used for satellite delivery. This implies significant extra complexity and it is likely that in some cases the cable operators will come to an arrangement with the satellite broadcasters over use of their conditional access control systems. The main markets for digital cable TV in Europe are currently seen to be in France and Germany.

4.4.2.4   SMATV

The DVB satellite master antenna TV (SMATV) specification which has also now been standardized by ETSI, includes two options [16]. For the first, the satellite signals are transmitted via coaxial cables at the intermediate frequency. This allows the use of standard DVB satellite TV set-top boxes. In the second case, the signals are transmultiplexed to the cable TV format allowing the use of standard DVB cable set-top boxes. The equalization requirements for the cables are defined and for new installations it should be possible to use the 64 QAM cable TV mode. However, for older installations the lower-rate options may have to be used. This would imply increased complexity and cost, not least because the existing amplifiers would have to be replaced, and so some networks may stay with analogue until they are upgraded.

4.4.2.5   MVDS

DVB has also defined specifications for microwave video distribution and these have been passed to ETSI. This approach is seen as a cost-effective way to increase the reach of cable TV networks in rural areas. As for SMATV, there are two options based on the existing satellite and cable specifications and boxes.

4.5   ADDITION OF NETWORK INTERACTIVITY

The Interactive Services Commercial Module (ISCM) was set up following the DVB decision to include interactivity in its specifications [7]. Two *ad hoc* groups were also set up within the Technical Module to define:

- protocols for interactive services (ideally independently of the delivery system) to implement the user requirements defined by the ISCM;

- modulation and coding for the complete range of interaction channels associated with the delivery systems.

4.5.1   Interactive services

The ISCM has defined commercial requirements for asymmetric interactive services supporting broadcast to the home with narrowband return channels comparable with those offered by PSTN/ISDN [17]. The following types of services could be supported by a set-top box (STB) meeting these requirements. The required level of functionality increases in each case.

4.5.1.1   Local interaction

In this case, data or software is downloaded to the STB and consumers are able to decide what the box will select and process (no interaction path is required). Also, software in the STB can be used to control other local devices (e.g. VCR, CD-ROM, CD-I).

4.5.1.2   Response to broadcast services

In this case the user can respond to broadcast material in an anonymous manner, for instance by voting in a talent contest. Another example is opinion polling. These cases only require the number of votes to be counted. The response can be immediate or by store-and-forward techniques.

4.5.1.3   Request for information

In this case the user requests information from a remote database with, for instance, details of a sporting event or other ongoing broadcast service. The data which is delivered can be conditional-access controlled. Once again this case can be implemented using immediate delivery or by store-and-forward techniques.

4.5.1.4   Purchase request — under broadcast conditional-access control

In this case the user purchases access to a broadcast service. This enables impulse pay per view where either tokens are downloaded to the STB or the services are authorized immediately following the request.

4.5.1.5   Purchase request — independent of broadcast conditional-access control

This case covers the more general requirement for financial transactions associated with, for instance, shopping, banking and gambling. It is necessary to prove the identity of the user rather than the STB and to confirm financial status. Delivery can be either over-air (e.g. software) or by mail (e.g. a suit or dress). Once again immediate or store-and-forward options can apply.

4.5.1.6   Messaging and remote monitoring

This case covers three types of service. Firstly, the user can initiate a message to the service or network provider (e.g. help). Secondly, the service provider can send messages to users (e.g. there is a new software release) or to the network provider (e.g. return channel is not working). Thirdly, the network provider can send messages to the user's STB (e.g. send viewing statistics at a particular time) or to the service provider (e.g. the following network contention measures have been taken).

Examples of some possible services are given in Mills and Dobbie [18].

4.5.2   Implications of commercial requirements

The following summarizes the ISCM commercial requirements in terms of their implications for specifications. The core requirements are listed first, followed by the optional requirements:

- a suitable data modem for all the networks to be considered (PSTN/ISDN and cable first) (note that where the network is shared with telephony services, it is necessary to detect when a telephone elsewhere in the home requires the line to allow for emergencies);

- protocols allowing for delivery of text, still pictures, graphics, audio, video and data including e-mail via the interaction channel — the protocols must allow for future network or service extensions and provide a means of:

— delivering information on server addresses, etc, to allow the STB to set up an interaction channel automatically, if authorised;

— informing the applications layer of relevant network-related information, including any network congestion measures;

— describing the capabilities of an STB allowing for future extensions;

— describing the services available so that the STB can identify whether a service can be accessed;

- flexibility to access a number of different servers through a range of different networks;

- a standard means of dealing with calls to STBs;

- optionally, a common, reliable downloading mechanism for new applications;

- optionally, a means of monitoring incoming telephone calls using Calling Line Identification;

- optionally, a protocol allowing remote checking of STBs for diagnostic purposes;

- optionally, a standardized interface allowing the addition of plug-in network interface modules.

4.5.3   Specifications in preparation

DVB is currently defining specifications to meet the ISCM commercial requirements.

4.5.3.1   Network-independent layers

The Systems for Interactive Services *ad hoc* group within the DVB Technical Module is close to completion of a specification covering the network-independent protocols, up to layer 4 on the OSI stack [19]. The implications of the specification for provision of interactive services are described in a guidelines document [20]. This also explains the terminology used for the logical signal flows and summarizes the functionality of the protocols identified in the specification.

Commonality with the Digital Audiovisual Council (DAVIC) protocol specifications [21] has been a DVB target and the current specification is effectively a sub-set with some minor differences to minimize complexity and reduce costs.

The specification is currently based on the use of user datagram protocol (UDP) or transmission control protocol (TCP) with Internet protocols (IP), and Moving Picture Experts Group 2 (MPEG-2) (Private Sections using the Digital Storage Media), Command and Control (DCM-CC) protocol (which was recently specified by MPEG [22]), sections for carriage of content (audio, video, data) via the broadcast channel.

The interaction channel uses UDP and IP with the point-to-point protocol (PPP) for time-sensitive (synchronized) content and TCP with IP and PPP for non-synchronized content. The multilink point-to-point protocol (MP) is used with PPP in both cases (e.g. for ISDN).

Downloading of data across the broadcast channel is possible using DSM-CC data carousels user-to-user interaction across the broadcast channel for application control, and communication is achieved using DSM-CC user-to-user (U-U) and object carousels. In both cases these are via DSM-CC sections (MPEG-2 Private Sections) within the MPEG-2 transport stream.

For the interaction channel, downloading of data is possible using DSM-CC download with TCP/IP/PPP(MP). User-to-user interaction for application control and communication is achieved using DSM-CC U-U with universal networked object (UNO) common data representation (CDR) and ONO remote procedure call (RPC) with TCP/IP/PPP(MP).

4.5.3.2   Network-dependent layers

The broadcast channel can be via any of the options specified by DVB as listed in section 4.4.2. In all cases the broadcast channel is used to deliver an MPEG-2 transport stream.

The open systems interconnection (OSI) physical layers for different interaction network options and the network-dependent protocols within the transport layer are being specified by the Return Channel *ad hoc* group within the DVB Technical Module. A specification for public switched telephone networks and integrated services digital networks (PSTN/ISDN) is close to completion [23] and specifications for CATV, SMATV and MMDS networks are currently being prepared (in that order or priority).

4.6   CONCLUSIONS

Broadcast technology is developing rapidly and the change from analogue to digital is almost upon us. It is likely that satellite services will be introduced first within Europe, but terrestrial digital TV services will follow rapidly, at least within the UK. The terrestrial services have the advantage that a dish is not required and that an enhanced package of 'free-to-air' services are likely to be provided in addition to new subscription services. In both cases it is likely that network interaction via the PSTN (or possibly the ISDN) will be offered as an integral part of the new services.

Specifications for the core processing and delivery of the broadcast services have been produced by the European DVB project and these are being standardized. These include a specification covering network-independent protocols for interactive services which is effectively a modified subset of the DAVIC specification for enhanced broadcast services and specifications for the most practical interactive channel network options.

APPENDIX

List of Acronyms

ADSL	asymmetric digital subscriber loop
ATM	asynchronous transfer mode
CENELEC	European Committee for Electrotechnical Standardization
COFDM	coded orthogonal frequency division multiplex
DAVIC	Digital Audio-Visual Council
DVB	European Digital Video Broadcasting project
DVC	digital video cassette
ETSI	European Telecommunications Standards Institute
FDM	frequency division multiplex
FM	frequency modulation
IEC	International Electrotechnical Commission
ISO	International Organisation for Standardization
ITU	International Telecommunications Union
MVDS	microwave video distribution system
PCMCIA	Personal Computer Memory Card International Association
QAM	quadrature amplitude modulation
QPSK	quaternary phase shift keying
SMATV	satellite master antenna TV
STB	set-top box
VSB/AM	vestigial sideband amplitude modulation

REFERENCES

1. Pilgrim M *et al*: 'The M³VDS Saxmundham Demonstrator — multichannel TV by MM waves', BT Technol J, 7, No 1, pp 5-19 (January 1989).

2. ISO/IEC DIS 13818 - 1: 'Information Technology — Generic Coding of Moving Pictures and Associated Audio Information Part 1: Systems'.

3. ISO/IEC DIS 13818 - 2: 'Information Technology — Generic Coding of Moving Pictures and Associated Audio Information Part 2: Video'.

4. ISO/IEC DIS 13818 - 3: 'Information Technology — Generic Coding of Moving Pictures and Associated Audio Information Part 3: Audio'.

5. Reimers U: 'The European Project on Digital Video Broadcasting — Achievements and Current Status', International Broadcasting Convention, IEE Conference Publication No. 397, pp 550-556 (September 1994).

6. Woods D: 'On the Eve of the Revolution — Digital Television Broadcasting in April 1994', EBU Technical Review, pp 3-12 (Summer 1994).

7. Digital Video Broadcasting Project Press Release: 'DVB now working on Interactive Services', Frankfurt, available from DVB Office, c/o EBU, Geneva, Switzerland (March 1995).

8. Windram M and Druru G: 'Satellite and Terrestrial Broadcasting — The Digital Solution', International Broadcasting Convention, IEE Conference Publication No 397, pp 366-371 (September 1994).

9. ETS 300 468: 'Digital Broadcasting Systems for Television, Sound and Data Services: Specification for Service Information (SI) in Digital Video Broadcasting (DVB) Systems', European Telecommunications Standards Institute, Sophia Antipolis (1994).

10. ETS 300 472: 'Digital Broadcasting Systems for Television: Specification for Conveying ITU-R System B Teletext in Digital Video Broadcasting Bit Streams', European Telecommunications Standards Institute, Sophia Antipolis (1994).

11. Digital Video Broadcasting Project: 'Common Interface Specification for Conditional Access and other Digital Video Broadcasting Decoder Applications', CENELEC (1994).

12. Digital Video Broadcasting Project: 'Interfaces for DVB-IRD', CENELEC (1995).

13. ETS 300 421: 'Framing Structure, Channel Coding and Modulation for 11/12 GHz Satellite Services', European Telecommunications Standards Institute, Sophia Antipolis (1994).

14. Comminetti M and Morello A: 'Direct-to-Home Digital Multi-Programme Television by Satellite', International Broadcasting Convention, IEE Conference Publication No 397, pp 358-365 (September 1994).

15. ETS 300 429: 'Framing Structure, Channel Coding and Modulation for Cable Systems', European Telecommunications Standards Institute, Sophia Antipolis (1994).

16. ETS 300 473: 'Digital Broadcasting Systems for Television, Sound and Data Services: Satellite Master Antenna Television (SMATV) Distribution Systems', European Telecommunications Standards Institute, Sophia Antipolis (1994).

17. Digital Video Broadcasting Project: 'Commercial Requirements for Asymmetric Interactive Services Supporting Broadcast to the Home with Narrowband Return Channels', DVB Doc No. A008 (October 1995).

18. Mills G S and Dobbie W H: 'DVB specifications for broadcast-related interactive TV services', Int Broadcasting Convention, IEE Conf Publication No 428 (1996).

19. European Digital Video Broadcasting Project: 'Network Independent Protocols for Interactive Services', Specification currently in preparation.

20. European Digital Video Broadcasting Project: 'Guidelines for the use of the DVB Specification: Network Independent Protocols for Interactive Services', Currently in preparation.

21. Digital Audiovisual Council (DAVIC): 'High and Mid Layer Protocols (Technical Specification)', DAVIC 1.0 Specification Part 07 (January 1996).

22. International Standards Organisation: 'ISO/IEC Draft International Standard 13818-6 MPEG-2 DSM-CC Specification', (December 1995).

23. European Digital Video Broadcasting Project: 'DVB Return Channel through PSTN/ISDN', Specification currently in preparation.

5

SERVERS FOR BT'S INTERACTIVE TV SERVICES

G W Kerr

5.1   INTRODUCTION

The task of storing, managing, and spooling out information for BT interactive TV services is not trivial, once reasonable numbers of customers are using the services — in the Market Trial (1995) system, 2000 hours of audiovisual material required nearly 2 Tbyte of on-line storage, with a potential of 1250 independently controlled 2-Mbit/s streams (a total of 2.5 Gbit/s) from the one server.

The majority of envisaged interactive TV services will be based on stored information, on a server or set of servers, and this chapter looks at various ways of meeting the requirements for various types of services that might be eventually bracketed under the umbrella of 'Interactive TV'.

5.2   WHAT IS A SERVER?

A typical server (see Fig. 5.1) consists of four major logical parts:

- audiovisual database, where the audiovisual files, and possibly their associated embedded control and graphics information, are stored;

- audiovisual information processor and pump, where the information to be spooled to customers is extracted from the audiovisual database, formatted where required, and sent to the network interface for transmission to the user — in addition, this entity would probably handle the bit-level loading of information from a live feed, or a mass-storage device, under the control of the server;

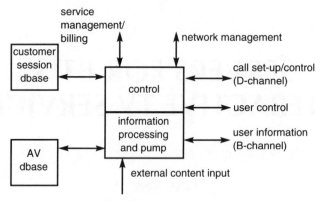

Fig. 5.1   A typical server.

- server control, which manages the server, its sessions, and its databases, and provides the intelligence to handle each user session, including session requests and clear down, while at the same time providing the intelligence to handle requests from the network management systems and service management systems, and packaging information for passing to the billing system;

- the customer session database, which is used by the server control to handle each user on an individual basis according to the various parameters dictated by the overall service and its status, the specific programme, and the individual users' preferences.

A server can be considered to have the following interfaces:

- stored information sent to user (and in some situations received from users);

- control channel to/from users, so they can control their sessions;

- call set-up/control to/from network;

- network management, so status and integrity of server can be remotely managed by the server provider;

- service management/billing, to permit the service provider to enable and disable services and customers' access to them, extract billing and use this information, and control the overall service provided;

- external content input, where new material can be loaded on to the server under the control of the service provider — this material might additionally or exclusively be available in real time as a broadcast feed, where permitted; the material will, in general, include audiovisual information, graphics and navigation/control information.

5.3   SERVER PARAMETERS

Various main requirements on a server will tend to indicate different technical solutions. The major requirements (see Fig. 5.2) are considered to be:

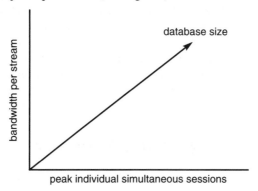

Fig. 5.2   The major requirements on a server.

In addition, the level of interactivity for each session will also have a bearing on the solutions provided.

5.3.1   Bandwidth per stream

5.3.1.1   Low bit rate (LBR)

LBR is taken to mean each stream is of bandwidth $n \times 64$ Kbit/s, where $n = 1$-6. Such servers operating at these speeds already exist for speech applications in the BT network, and can also be readily used for facsimile databases and telemetry systems. With few changes, such servers could also be used for database services for ISDN videophone and ISDN multimedia terminals, such as the PC-based VC8000. The LBR servers can interface directly to the BT transmission network either as a peripheral (for example, connected to a series of standard ISDN lines) or as part of the network (30-channel multiplex and CCITT No. 7 signalling).

Such servers could be used as part of a BT interactive TV service where:

- LBR audiovisual quality is acceptable through the use of still images, graphics or carefully chosen material content;

- or the server acts as a controller of the customer session and the audiovisual signal is transmitted via another system.

5.3.1.2   Broadband (HBR)

Broadband is taken here to mean where each stream is $n \times 1$ Mbit/s, with $n = 2$-8. This bit rate can provide entertainment quality TV (or better), but at present widely deployed networks do not exist which can handle such streams to users. Other chapters in this book address the issues of broadband connections to users.

The BT Interactive TV Market Trial (1995), and the earlier product development trial (1994), both used broadband streams at a total bit rate of 2.048 Mbit/s per stream. This bit rate is considered to give sufficiently good quality for most film and TV material which has been encoded off-line, but some customers may wish to pay more for EDTV or even HDTV quality in the future.

5.3.2   Peak traffic (sessions)

As far as the server is concerned, this parameter relates to the peak simultaneous independent sessions that it is required to support. BT is currently only permitted to carry fully one-to-one entertainment services over its network, so, in that situation, one independent stream corresponds to one user. In systems used mostly for transactional services, this is not a real restriction, but, for systems focusing on the film/TV programmes services, some bunching of users might be technically possible in order to reduce costs — this bunching is variously known as narrowcasting, or simulcasting, and the benefit to the provider is an increase in users per server stream.

For a given performance, the parameter of peak simultaneous sessions has a major impact on the server costs, as it affects:

- network interfaces, as the larger the number of peak simultaneous sessions at a given bit rate, the higher the capacity of the network interface required;

- bit-carrying power within the audiovisual information processor and pump;

- organization of storage, as a higher traffic requirement will mean a greater aggregate sustained data rate from the overall audiovisual storage system;

- performance of the server control sub-system, as more processing power will be required to provide a given quality of service to users as the number of peak sessions increases.

It is therefore not surprising that early servers from suppliers have tended to provide a maximum of 100-200 simultaneous sessions at ~ 2 Mbit/s per stream, because, beyond that, special design of some of the various server sub-systems is required.

5.3.3   Audiovisual database size

For a given bit rate per stream, the database size (in gigabytes) clearly is related directly to the amount of material to be stored (in terms of hours of material). Conversely, the reverse is also obvious — for a given number of hours of material, the size of the database (in gigabytes) is directly related to the bit rate of the streams to users. However, the amount of raw-data storage required is also dependent on implementation details, such as:

- replication of information in order to meet the requirement for the peak number of sessions to be supported;

- use of derived material from the raw source in order to reduce real time processing needs, e.g. visible fast-forward or fast-reverse files;

- use of redundant information in order to protect the on-line data from most likely storage media failures;

- storage fragmentation, as a result of material being updated or churned;

- buffer storage in order to carry out storage defragmentation;

- use of space to store new material prior to the old material being deleted and the new material being made available to users.

Various technologies can provide cost-effective solutions for the different service scenarios, and these will be discussed later in the chapter. While considering the design of the database system, the trends in cost/performance of raw data storage for various storage media need to be considered carefully.

5.3.4   Interactivity

Little and Venkatesh [1] categorize interactive services into the following five groupings, based on the amount of interactivity allowed:

- broadcast (no video-on-demand (VoD)) services similar to broadcast TV, in which the user is a passive participant and has no control over the session;

- pay-per-view (PPV) services in which the user signs up and pays for specific programming, similar to existing CATV PPV services;

- quasi-video-on-demand (Q-VoD) services, in which users are grouped based on a threshold of interest — users can perform rudimentary temporal control activities by switching to a different group;

- near-video-on-demand (N-VoD) services, in which functions like forward and reverse are simulated by transitions in discrete time intervals (of the order of five minutes) this capability can be provided by multiple channels with the same programming skewed in time;

- true video-on-demand (T-VoD) services, in which the user has complete control over the session presentation, having full-function VCR (virtual VCR) capabilities, including forward and reverse play, freeze, and random positioning — T-VoD needs only a single channel (multiple channels become redundant) and is sometimes called interactive VoD (I-VoD).

T-VoD services can probably also be subdivided into:

- films on demand — users choose the item of interest and then control their viewing as if on their local VCR;

- browsing and transactional services, such as home shopping.

N-VoD is also sometimes called 'staggercast'.

In the one extreme case of broadcast, the interactivity level is zero, so the demands on the server are negligible; at the other extreme of transactional services, the demands on the server will be similar to that on a public database service, and the processing requirements for such servers, when dealing with many thousands of simultaneous sessions, are considerable. The demands of interactivity will mostly affect the server control, and so will tend to be independent of bit stream bandwidth to the user.

5.3.5   Other server parameters — server sizing and location

An overall service to users does not necessarily have to be provided by a single server platform, and in many situations it may be more cost effective to use a multiplicity of platforms, especially where the users are scattered over a large geographical area. In addition, for a company such as BT, there is the option to consider the servers as being part of the telecommunications network, or simply as attached peripherals to the network. These parameters are considered here.

5.3.5.1   Network-based servers

Network-based servers can either be provided on the basis of a few large centralized servers for a large geographical area (for example, the UK), or many small and localized servers serving smaller numbers of users each. Both approaches have their advantages, and will be discussed below.

When considering network-based servers, in order to reduce operational costs it is vital that they have high availability, and can be remotely managed to the network management requirements of the service provider. In addition, they will probably need to provide a standard interface to the service provider's service management and billing systems. Servers in the network can be configured to be 'intelligent peripherals' as part of the intelligent network, or a formal attachment to the network and yet reside on the service provider's premises. Low bit rate servers can use existing signalling systems (such as CCITT No. 7) and so fit into the existing network structure; broadband servers will have to meet emerging signalling standards, such as the possibility of the ITU-T Recommendation Q.2931 as a peripheral.

In large systems, it has been suggested that servers be considered in a hierarchy, depending on the demands of users on their material; a more local server can pass on the request to a remote server if the information the user is requesting is not available. It needs to be noted that this approach can make considerable demands on the local server to handle the information from the remote site, and current modelling would indicate that this may not be cost effective. However, different systems and services will demand different solutions. Irrespective of the number of servers provided, each one will need to be kept in date and in track with other servers, especially where material is replicated across servers. With a small number of large servers, the information tracking can be a fairly straightforward affair, but this does mean that high-bandwidth links are required from local 'points of presence' all the way through to these servers, with their associated costs, whereas when using smaller servers at local points of presence (for example the local exchange) the bandwidth of the main network to some form of 'update' server is going to depend entirely on the rate of update of information required, leading to some potential savings in transmission. The effects of having servers in local exchange buildings must not be underestimated, as the operational costs of providing maintenance cover across so many locations UK-wide will inevitably be large in order to provide an acceptable quality of service to end customers. Figures 5.3 and 5.4 show the two main options, although others obviously exist.

It must also be noted that when opting for the use of smaller local servers, there may well be greater up-front costs at the early stages of service deployment, as a completely functioning server is required at each local exchange where the first user is connected, whereas, with a remote larger server, that server can be built up as the overall traffic increases. The main operational costs as far as the server is concerned will focus on support for the storage media, as these tend to involve mechanical items, and do not offer high reliability compared with all-electronic systems.

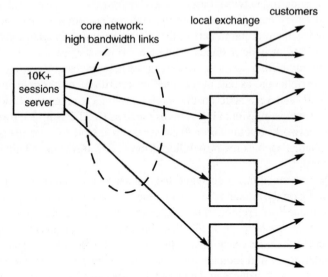

Fig. 5.3   Large server within network.

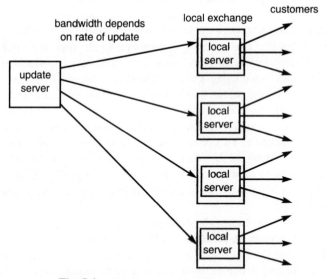

Fig. 5.4   Local servers updated from remote system.

5.3.5.2   Customer-based servers

These servers tend to be smaller units, and are only 'attachments' to a network. They only require interfaces to meet the service requirements of the service provider, and will probably not be remotely managed, and so will not be part of any intelligent network or network management system. However, various solutions are already emerging for such smaller systems for the hotel, call centre, and airline entertainment markets.

5.4   BROADBAND SERVERS — SOME OF THE FAMILY

Broadband servers tend to break down into various categories, depending on the level of interactivity required and the scale of the server, for example:

- large-scale, interactive for transactional multimedia services (10K-plus peak sessions);

- large-scale for TV programmes, films, interactive VoD only;

- large-scale for N-VoD type services;

- trial cable-TV and airline/hotel types (~ 200 sessions).

5.4.1   Interactive broadband server

A typical interactive broadband server for a large network system (see Fig. 5.5) will probably address something like:

- 10K to 50K-plus simultaneous sessions;

- greater than 20 Gbit/s downstream capacity total;`

- greater than 100 Mbit/s user control peak capacity;

- 2000-plus hours of material — greater than 2 Tbyte audiovisual store.

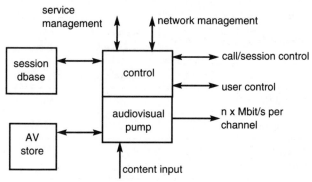

Fig. 5.5   A typical interactive broadband server.

In contrast, a customer-based system will probably demand:

- 200 simultaneous sessions;

- greater than 400 Mbit/s downstream capacity;

- greater than 2 Mbit/s user control peak capacity;

- 200 hours of material — greater than 200 Gbyte audiovisual store.

The server control will depend on the level of interactivity of the service required — if the bulk of customers will be using transactional services, then the demands on the control part will be far greater than those just used for films and TV programmes on demand.

5.4.1.1   Audiovisual store implementation

For an interactive broadband server, some material will be very popular at a particular time, other material will be rarely accessed, but, in most systems, users will demand that material is available 'instantly'. This would tend to imply that slow media, such as tape, are not feasible for this service, especially if tapes are loaded mechanically on to a tape reader from a demand of a user (response times liable to approach 0.5-2 minutes). However, some implementors consider one solution to be to store the first few minutes of each title on a rapid-access storage medium (like magnetic disk) to cover the time taken to load the tape; others state that the cost/bandwidth of tape is not much different from disk once all aspects are considered, and focus entirely on disk-based systems. Some implementors believe that solid state (RAM) storage for popular items can reduce overall costs. Table 5.1 contains data from Little and Ventkatesh [1] regarding storage media.

Table 5.1   Comparative storage media costs [1].

Storage Type	Cost/Mbyte	Sessions/ Device	Cost/Movies stored
RAM	$50.00	200	$50,000
Hard disk	$0.50	5	$400
R/W optical	$0.20	2	$200
Magnetic tape	$0.01	1	$10

Note: These are 1994 indicative costs, and, as the market is so changeable, the figures can only be taken to show rough comparisons.

From the data in Table 5.1, comparative sessional costs can be derived (see Table 5.2).

Table 5.2   Comparative sessional storage media costs.

Storage Type	Cost/Session/ Mbyte	Cost/Session/Film
RAM	$0.25	$250
Hard Disk	$0.1	$100
R/W Optical	$0.1	$100
Magnetic Tape	$0.01	$10

However, the tape and optical disk options exclude (presumably) the cost of the readers and any mechanical handlers, and so perhaps are not directly comparable with the others.

The majority of solutions to date for this type of broadband server have focused on the clever use of magnetic disk stores, either as one complete entity, or spread across multiple processor or spooling elements. In any sizeable disk store, the design needs to take into account maintaining the data integrity in the event of a failure of an individual disk drive, as the MTBF of disks is sufficiently finite to make failure of one drive in a large system quite a probable event over a period of months — this is not surprising, given the mechanical nature of disk drives. Techniques such as RAID [2] can be used which enable a faulty disk drive to be replaced and the data automatically rebuilt on to it, although clearly this is only possible where the original data has been stored with sufficient additional redundancy to enable the lost data to be derived without reloading from back-up.

In some scenarios, it is very hard to predict the demand patterns across the various parts of the database (for example, the popularity of a specific film or presentation), and the database then has to be so designed to meet these unknowns. One such technique is 'disk striping', where a given audiovisual item is scattered across a large number of disk drives, so a given user is constantly and invisibly moved from drive to drive while watching one item; each drive is filled with segments from many different audiovisual items, so the probability of demand on a drive can be fairly well defined. Clearly such a system demands very careful data management, and also the necessary algorithms to handle demand appropriately as it arrives.

In addition, in a number of services, material will need to be updated, old material deleted and new material loaded. This 'churn' process demands careful management of the database, in terms of:

- efficient use of database capacity, while providing continuous service to users;

- careful management of data bandwidths so that storing new material does not saturate the capabilities of the database system to provide service to users.

Turning to the processor elements that control and spool the data from the disks stored to the network interface, these elements need to be able to handle data very rapidly, but usually without performing additional processing. Suppliers to date are tending to focus on either producing bespoke hardware to perform this function, modifying some form of 'massively parallel processing' (MPP) system, or have a multiplicity of standard, smaller-scale processor systems with their own associated disks, and switching their stream outputs.

Proponents of MPP systems claim the following advantages:

- the CPU elements are mass produced, and therefore low cost — in some cases the MPP vendors use standard CPUs, so do not pay directly for the development of new generations of CPU, but use them as they become available;

- MPP architectures can inherently offer large overall bandwidths, as any one path does not carry large amounts of traffic;

- spare processor power can be used if required;

- expansion in terms of database size and/or simultaneous users is easily performed without major structural impact;

- no additional switching is required on the network side, as all switching is performed within the MPP system, so any stream can appear on any physical network interface.

Proponents of the bespoke solution would claim the following advantages:

- systems specifically designed to meet server requirements;

- no wasted system or processor power;

- more readily reduced in size, power consumption and cost as the market size increases and technology permits.

Proponents of the multiple small-scale processor systems tend to opt for PC (or equivalent) systems, claiming advantages as:

- PCs are commodity items and very low cost;

- expansion is relatively easy — you buy more PCs;

- very little hardware development cost.

The main drawback of some proposed systems lies in the need (or otherwise) of the server to supply any session on any system-defined network interface, not necessarily the interface that would be optimum for the server. To that end, some designs include a captive switch, especially where ATM-based network interfaces are used.

5.4.1.2   Server control

In many ways the control part of the server has opposite requirements compared with the audiovisual control — as the server controller is required to handle all the interactions with the customer, and control the overall server, it needs processing power and intelligence, with less emphasis on simple data handling.

Many vendors to date are therefore focusing on providing high-performance Unix™-based systems of a size dependent on processing power required. These systems are either standard Unix workstations, or can be symmetric multiprocessing (SMP) or even MPP systems.

Clearly in some scenarios, an MPP vendor might choose to integrate the audiovisual pump processor/spooler with a server control, allocating processors with an overall MPP for the server control, some for the network interface, and some for the spooling activity; other vendors might choose to allocate processors on a more dynamic basis as instantaneous demand dictates. This, though, will require some highly sophisticated low-level software.

The main focus of the server control must be to handle large numbers of user (and associated call-control) interactions with a good overall response time. Much of the user's perception of response time will depend on the overall software architecture used and hence the processing performance of the set-top box (see Chapter 6), but some of this processing is inevitably performed by the server, however sophisticated the set-top box. Again, careful sizing of the server control to meet the needs of peak simultaneous sessions, expected user interactions, and the software environment chosen, will be required if the user is to perceive a good quality of service, especially when using transactional services. Much of the sizing can be based on traditional sizing models for multi-user real-time systems, and is not discussed further here.

So far, the server model used has assumed a single, logical element for the server control; in some situations there may be the potential of having a set of loosely coupled processors controlling one audiovisual database system, each processor handling a set of user sessions. This may be of particular importance if the user applications are highly processor-intensive for the server.

5.4.1.3   Content loading

At first glance, the loading of encoded audiovisual information and associated data on to the server is a trivial file-transfer problem, but once consideration is given to the large amounts of data involved, it can be seen as in fact a non-trivial operation. Content can arrive in a number of ways:

- mass storage device, e.g. computer tape, containing a complete encoded audiovisual file to load;

- a communications link to a remote system, over which encoded audiovisual files are sent;

- a communications link from which a live or broadcast programme is received.

Common computer tape systems can spool at perhaps 6 Mbit/s, so for each hour of material encoded at 2 Mbit/s, approximately 20 min of loading time is required. An additional set-up time (to put the tape in the device, issue commands at the operator console, permit the tape to be checked by the processor before spooling, and at the end remove the tape) might be five minutes or more; therefore, assuming film/TV programmes of average duration 90 min, one tape drive might support the loading on average of 1.7 programmes/hour. If a server has 2000 hr of film/TV material, then at the start there might be 1333 programmes to load, which would take perhaps 780 hours to load on one tape drive. This corresponds to nearly 100 standard working days if one drive is used, with the server fully occupied performing this file loading.

In the situation of obtaining material over a communications link, much the same discussion applies as that from a tape, or other mass-storage device, except that more checking is possibly required to ensure the information arriving is that which is expected, and is dealt with appropriately. The bandwidth of the communications link required will directly relate to the rate of material update required; higher-level protocols to use when sending programme information to a server are only just emerging from industry associations such as DAVIC (Digital Audio-visual Council [3]. The server must also be sized to be able to accept and process the incoming information at the rate received in addition to continuing live service to users. Such processing may not be trivial if the local server is required, for example, to derive visible fast-forward and fast-reverse files from MPEG play files.

In the situation of content arriving from a live or broadcast programme feed, different criteria will apply. Broadcast feeds will clearly be one-way, and there will be no way the server can request flow control on the information, so it must be capable of receiving and dealing with the information as it arrives. In general, the information will be simply an encoded audiovisual stream, with no inherent content, graphics, user application, or menu data; these elements will have to be provided either separately or manually. As the broadcast feed may be received in unencoded form, a real-time encoder and multiplexer will be required with appropriate hardware interface to the server. Currently, such encoders are relatively expensive. The server may additionally be required to make the broadcast feed immediately available to users, either on a 'live' basis, or on a basis of their being able to watch from the start-up to the current point of the programme, with use of some VCR controls as appropriate. Clearly in the BT situation, the current regulatory environment prohibits an entertainment broadcast feed being made available to users, so currently only the option to store the audiovisual information is permitted. Some form of additional control information will be required from the broadcast feed, so the server is aware of the actual start/end of programme segments or items, if they are to be stored for further use.

However, the loading of the material is not simply the audiovisual content, but must also include the categorization of the material, menu routeing to that material, and text and graphical information on that material, so the eventual users can be directed to their choice. The actual mechanisms to be used for this are the responsibility of the applications software, but, while standards exist for audiovisual encoded files, they do not yet exist for all the ancillary information for films, and a lot of manual intervention is required. Usually this heavily increases the time taken to load fully a new programme on to the system.

While the problem of loading is perhaps most acute at service start-up, it does not disappear if there is to be a significant amount of programme churn on the server (i.e. new programmes being introduced and old ones being removed). As such churn will take place on the live server, it is vital that the process ensures that the integrity of service is maintained at all times to users, and that menus accurately reflect the content and programmes actually available to a user.

When transactional services are considered, additional information is required to be loaded, relating to the application-specific processing required by the service, whether that processing takes place in the server, set-top box, or on both. Much of the control information will be produced from an authoring package, by the information provider, but the audiovisual segments will probably need to be encoded separately and brought together with the other information to produce the complete service. A transactional service will therefore probably consist of many files, and the management of these while loading will be vital to the integrity of the service.

5.4.1.4   Content definition

Standards for the description and definition of files for server contents are currently being discussed in forums like DAVIC [3], but no decisions have been made on this at the time of writing. In general, contents being loaded on to a server will need to include definitions of:

- linear audiovisual segment (embedded audio with moving video, such as an MPEG multiplex) — parameters will be required to define encoding and multiplexing used;

- still-video image — parameters required for defining screen resolution and encoding system used (e.g. JPEG) and screen location/relative size;

- audio segment — parameters required to define encoding and any multiplexing used, and possibly, number of audio channels with the multiplex provided, and whether those channels are alternatives (e.g. multi-lingual) or multi-channel sound, or some mixture of the two;

- graphics object — parameters required to define encoding (and any multiplexing) used, intended relative/absolute screen position and overall size in relation to the overall screen size, whether overlaid or replacing any video and/or existing graphics on screen, whether any movement (time, vector) or change in shape (3-D vector and/or mesh definition) while displayed parameters are required to define the performance of additional software requirements (e.g. in the terminal);

- cursor/highlighting object — parameters to define when the cursor is used, its appearance (any graphics will need the definition of encoding supplied), any blinking rate, etc;

- execution instructions — parameters are required to define the condition(s) for the execution to occur and the commands which execute them;

- signal-state table — the complete set of actions required for all possible interactions (user, system, time) with the application at the application's current state (will point to executables and/or display objects);

- execution map — a definition of the start and end states for the application and the top-end management programme for the given application;

- mass-storage requirements — parameters will include the (perhaps contiguous) space required for the various objects within the content and the intended bit rate, if not implied from the encoded material object, so that sufficient mass-storage bandwidth can be allocated within the server;

- execution requirements when running application — parameters will include a definition of the assumed requirements from the authoring tool, so that the receiving server can ensure that it loads appropriate run-time environments for the given executable (either or both on server and the sets of set-top boxes potentially connected to it), and, at the time of execution, can ensure that adequate server resources are allocated to the application — some default sets of execution parameters would be expected to be given which will work to the emerging industry-standard authoring tools at current versions, and for these to be updated in an agreed manner as tools are further developed;

- minimum requirements of user interface input device, for example, remote control with DAVIC-defined minimal set of keys, mouse, QWERTY keyboard, speech-driven system, 486 PC, Mac, etc;

- external communications/facilities required to run applications, for example, access to EDI, X.400, X.25, ISDN, facsimile/PSTN, Internet (perhaps with recommended minimum rates so that the content providers can guarantee some quality of service for their applications);

- execution requirements for loading content — parameters will include any unpacking of files, any mapping of logicals to physicals, whether provided content includes, for example, any derived information such as visible FF or FR information to show fast forward/fast reverse of the linear audiovisual information;

- on content coming from a telecommunications link, objects are needed at some stage to define whether the material is to be made available immediately (broadcast), and if so, whether the material is also able to be stored for later retrieval by customers (purely transitory or not, and if not, for how long it can be retained), the encoding and multiplexing employed on the link, and whether the material is arriving in real time — on live links, the objects describing the material may well arrive on a separate physical or logical link, but that makes little difference to the definitions (there may well be specially defined 'live broadcast' parameters which are used by default, since a live broadcast, by definition, cannot be stopped and restarted from the top);

- definition of class of material, in terms of predefined groups able to use it — parameters might include (for entertainment) class of viewing certificate, payment band, availability by time of day and/or after a given release date and/or rendered unavailable after a specific date/time; for other material, parameters might include predefined CUG classes (e.g. schools, educational establishments, registered hospitals only);

- descriptive and menu information for the content, such that any picture icons, descriptive text, major names, subject matter, etc, can be automatically entered into search databases, and content navigation systems and menus on the server.

5.4.2   Small-scale interactive broadband servers

Smaller-scale broadband servers tend to be directed more towards the films/TV programmes-on-demand services, and so they can be focused much more closely on the small number of users or sessions, and the specific application for which they will be used. Applications are already appearing not only in trial cable-TV systems, but also in hotel systems and airline systems. The controlling software tends to be a lot simpler, so the audiovisual pump and controller tend to be integrated, and are based on a PC of some form and/or a workstation. Some systems are emerging from disk integrators, exploiting their own knowledge of the detailed performance parameters of magnetic disk systems.

5.4.3   Staggercast broadband servers

A staggercast server (see Fig. 5.6) works on the basis of broadcasting a small number of popular titles, with each title broadcast a number of times, each time staggered by, say, 5-10 min, so that a user, on requesting a title, simply has to wait for the next start time, up to the maximum stagger time used by the system. In such a system, users can 'pause' their viewing, and then take up viewing on the next delayed version of the same title which most closely matches their pause position, and can 'fast forward' and 'fast reverse' by browsing through the other broadcasts of the same title, but only in multiples of the stagger time. The service to users can only consist of a few popular titles, and, in general, real transactional services are not available.

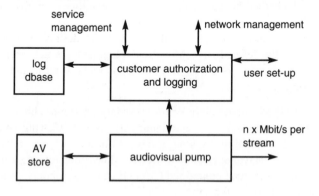

Fig. 5.6   A typical staggercast broadband server.

If a server is assumed to be offering 100 channels carrying 20 films, for a 2-Mbit/s system this would imply:

- approximately 200 Mbit/s downstream capability;

- 30 Gbyte on-line store.

This is clearly a much simpler task than the fully interactive broadband server, and the server needs to have little interactive capability, save managing user log-ins and billing, and the crude navigation outlined above.

Implementation of such a server can take various routes, including simply cutting down an interactive broadband server defined above. One alternative is to exploit the prior scheduling of the spooling activity and use storage media that are cost-effective for this, like tape. One type of tape system which can be of interest is that used for digital studios, namely D1 or D3. As these tapes provide user information at 270 Mbit/s, one machine could be used to spool out over 100 channels (at 2 Mbit/s) in parallel, for the length of the tape, which is usually 45 minutes at maximum.

Two machines can be readily synchronized so the second takes over from the first to provide a seamless join. In addition, as small- and large-scale mechanical tape-handling systems already exist for the broadcast station market, the loading can be entirely automated. Special systems will be required to master the original tapes (including all the staggering of the showings of the films), but this is not a major issue. Churning of the information will require new tapes being produced with the new correct mix of information. Such tape systems would only be loosely coupled to the server control to report status and control at a high level, programme tape loading, and timing. Broadcast tape machines are relatively expensive, but are designed for high availability and low maintenance cost, and few are needed for such a system.

5.5   LOW BIT RATE SERVERS

For completeness, a brief discussion on low bit rate (LBR) servers is included. As defined earlier, LBR servers aim to address services requiring $n \times 64$ kbit/s user streams, where $n = 1 - 6$.

Where $n = 1 - 2$, this can most easily be provided over standard ISDN and so can be used directly on BT's existing network; additional access network considerations may be required where n approaches 6 (384 kbit/s). A typical LBR server is depicted in Fig. 5.7.

The speech applications platform (SAP) already used in the BT network, is such a server, initially providing speech-based services (CallMinder, etc) [4]. Most such servers are based on a flexible architecture, so additional services can be added later by software changes and without changing the installed hardware system.

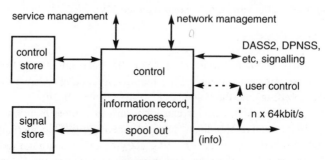

Fig. 5.7   An LBR server.

These types of platform are already being extended mostly by additional software, to provide other LBR services such as telemetry, facsimile, new charge-card-type services, as well as enhanced speech services. With some additional processing capability and software, they could also be enhanced to provide the audiovisual and multimedia equivalent of many of the above, including videophone database, videophone 'Callminder', VC8000 database dial-in, and dial-in audiovisual information services where these do not contravene the regulations regarding broadcast services.

As already mentioned in earlier sections, limited audiovisual services could be provided from such systems. At lower (ISDN2) bit rates, small-size moving images, full-screen graphics and still images can be used to good effect for a number of interactive multimedia services. Current telephony tariffing might restrict the use of such a system for long hold-time sessions from residential users, as it would appear that they would have to make at least a local call equivalent to one or two PSTN phone calls to access the server. The discussion on offering differential network tariffs for different services is not within the scope of this chapter.

Instead of providing audiovisual material, an alternative scenario could be to use such network-based LBR servers to provide authorization, perhaps some limited interactivity, and logging of events for services whose audiovisual element is provided over another means, such as digital satellite, cable-TV or digital terrestrial TV. In this situation the LBR server effectively replaces the server control part of a typical staggercast broadband server (see section 5.4), but uses the existing BT network for control and interactivity communications from (and possibly to) the customer.

5.6   CONCLUSIONS

There is unlikely to be one technical solution for all VoD server requirements, as these can vary considerably, dependent on overall service requirements.

Large-scale interactive broadband servers still present a considerable technical challenge to server suppliers, but technical solutions are emerging, and cost trends would indicate that commercially viable solutions will be possible in the near future.

Servers with less functionality than the fully interactive broadband server can be provided at considerably less cost per user stream, but the reduction in functionality may not be acceptable for all planned interactive TV services.

REFERENCES

1.   Little T D C and Venkatesh D: 'Prospects for interactive video-on-demand', IEEE Multimedia, 1, No 3, pp 14-21 (1994).

2.   The RAID Advisory Board: 'The RAIDBook: A Source Book for RAID Technology', Second Edition, 13 Marie Lane, St Peter, Minnesota 56082-9423, USA (November 1993).

3.   Chiariglione L: 'DAVIC — the Digital AudioVisual Council', in Proc of Telecom 95, Geneva (1995), also http://www.davic.org

4.   Rose K R and Hughes P M: 'Speech systems', in Dufour I G (Ed): 'Network Intelligence', Chapman & Hall, pp 75-94 (1996).

6

THE SET-TOP BOX FOR INTERACTIVE SERVICES

R A Bissell and A Eales

6.1 INTRODUCTION

The definition of a set-top box (STB), for the purposes of this chapter, is an item of consumer electronics which interfaces with a TV to provide an additional service.

The earliest high volume digital STBs were probably the first electronic 'tennis' games which emerged in the early 1970s, allowing one or two players to play a game displayed on the TV screen. The most common STBs currently (in the UK market) are video cassette recorders (VCRs), followed by analogue satellite receivers and cable boxes. The cable STBs are currently differentiated by the fact that, almost without exception, they can only be rented and there is little or no choice available to the customer.

The major piece of enabling technology for the emerging digital STBs has been the development of the MPEG (Moving Picture Experts Group) standards [1]. This has enabled low-cost digital decoder ICs to be developed by a number of silicon vendors to ensure healthy competition in the market-place.

Games consoles have grown in complexity, many with CDi™ or CD-ROM drives, and are today the most common digital STB. Digital satellite receivers have been developed to the emerging DVB (European Digital Video Broadcasting project) standards (see Chapter 4), service being due to start in the UK in late 1997. Digital satellite services have already commenced elsewhere in the world.

It must be remembered that the success of STBs for any new service will be totally dependent on the amount and quality of the content available. The market is therefore 'content driven'. It is essential to standardize the content format to provide stability to the consumer electronics manufacturers. For this reason many consumer electronics manufacturers have invested in content production facilities.

6.2   BT INTERACTIVE TV TRIALS

6.2.1   Overview of trials

The Interactive TV technology trial which commenced at Kesgrave, Suffolk, March 1994, was the first truly interactive video-on-demand (VoD) trial world-wide. The STB for this trial was supplied by Apple Computers Inc (Apple) and was based on a Macintosh computer with an additional 'motherboard' containing the line interface, MPEG decoder, graphics overlay and PAL encoder. The product was developed by Apple in the four months leading up to the trial and was housed in an 'off-the-shelf' metal rack enclosure, with a 'universal' infra-red remote control with custom screen-printed key labels.

The STB was redesigned for the Interactive TV market trial which commenced in the autumn of 1995. A new consumer electronic housing was designed, complete with a new infra-red remote handset. The Interactive TV market trial also provided, in addition to video on demand, many other services including home shopping, home banking, community information, education services and video games download.

The technology trial STBs each had to be manually configured prior to installation at the customer's premises. While this was acceptable for a trial of 60 BT employees, processes were put in place to ensure that the market trial STBs, delivered to 2500 customers, were self-configuring on installation.

6.2.2   Architecture

To position the STB in the overall Interactive TV architecture (see Fig. 6.1), it can be considered as the client in a client/server system, with the server being located in an exchange building and transmission being via either ADSL (asymmetric digital subscriber line) [2] or APON (ATM passive optical network) fibre to the home.

6.2.3   STB overview

The function of the STB is conceptually simple — it acts as a data termination device, converting digitally encoded audio and video to a more readily displayable form (e.g. PAL). Locally generated graphics are overlaid upon the video as required, producing a multimedia display. Within the structure of the Interactive TV trial architecture, the STB provides the 'client' functionality.

The STB provides a focus for the interactive element of a service, interpreting and transmitting a user's commands back to the 'server', e.g. 'fast forward', 'pause', 'rewind', and so on.

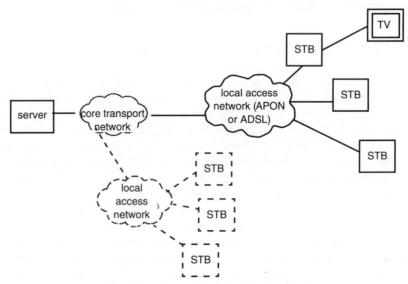

Fig. 6.1   Simple Interactive TV trial architecture.

The set-top boxes chosen for the Interactive TV trials are 'intelligent' systems based around standard desktop computers from Apple. In developing these units the approach was to leverage Apple's (industry standard) QuickTime™ media software layer, thus providing cost savings, and also some guarantee of design reliability. The additional functionality, i.e. that required by an STB over and above a normal desktop computer, was added by way of peripheral hardware (and associated device software).

6.2.3.1   STB details

The Apple Macintosh LC475 was chosen as the basis of the STBs because of its strong price/performance characteristics.

The Apple Interactive Television Box (ITV Box) exists in two versions — the 'prototype' model (STB1), as used in the 'Interactive TV technology trial', and the production model (STB3), as used in the 'Interactive TV market trial'. While both of these are based around the same core technology, experience gained from the development and use of the STB1 led to a number of changes in the later developed STB3. The capabilities of these two STBs are summarized in Table 6.1. Figures 6.2 and 6.3 depict the STB1 and STB3, respectively. (Apple considered the development of an STB2; however, this was an interim prototype, which did not go into production).

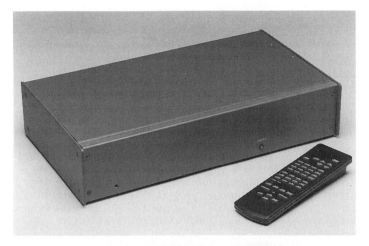

Fig. 6.2   Technology trial set-top box.

Fig. 6.3   Market trial set-top box.

The changes between these two set-top box variants are due to a number of factors, some technical and some commercial. The major changes in design policy were:

- the need for increased graphical performance (the market trial providing a multitude of extra multimedia services, against the single video-on-demand service offered on the technology trial);

- moving to a standardized MPEG2 transport layer;

- the 'real customer' needs of production engineering the set-top box into a practical and acceptable enclosure for use by the general public in their homes.

Table 6.1   Interactive TV trial set-top box characteristics.

Capability	STB1	STB3	Notes
Base System Microprocessor RAM data interfaces	LC475 68LC040 25 MHz 4 Mbytes none*	LC475 68LC040 25 MHz 4 Mbytes SCSI, RS422/232, Apple Desktop Bus	expandable to 36 Mbytes * STB3 options were available internally to the STB1
Video Capabilities MPEG decoder number of display planes graphics plane resolution video output	 C-Cube CL450 2 640 × 524, 8-bit colour SCART/UHF	 C-Cube CK450 2 640 × 524, 16-bit colour SCART/UHF/ S-video	MPEG 1 video 1 × MPEG video 1 × local graphics see Fig. 6.6 for overlay details
Audio Capabilities	stereo	stereo	
MPEG Transport Stream	proprietary framing	MPEG2 transport stream	
Network	G.703@ 2.048 Mbit/s RS232 @ 9.6 kbit/s	G.703 @ 2.048 Mbit/s RS422 @ 9.6 kbit/s	G.703 undirectional (broadband channel to STB) RS422/232 bi-directional (signalling channel to/from STB)
Physical Design	2 PCBs in 'lab case'	single cost reduced PCB in industrial designed case	
BT Production Requirements	100	2800	approximate figures to date.

The STB1 circuitry was also cost reduced to produce the STB3, with redundant LC475 functionality being removed and some general performance improvements made.

6.2.3.2   Hardware and software architectures

The **hardware architecture** of the Apple ITV boxes can best be described as a 'CPU-centric' design. All data flows via the 68040 processor, e.g. MPEG data received from the broadband data port passes via the processor, where the MPEG transport stream is software demultiplexed into audio, video and private data (non audio/video data, e.g. program specific information) to the MPEG decoder module where it is processed. While this design places a high load upon the processor's resources, it benefits from closely matching Apple's 'software-centric' approach to systems design. A further advantage, especially for the trial units, is that much of the architecture of the unit can be modified without need for major alterations in the unit's hardware design. Much of this flexibility is achieved through the use of a field programmable gate array (FPGA). The FPGA is used to provide the 'glue logic' required to interface the LC475 to the other modules within the set-top box.

The major hardware functionality within an Interactive TV set-top box is shown in Fig. 6.4.

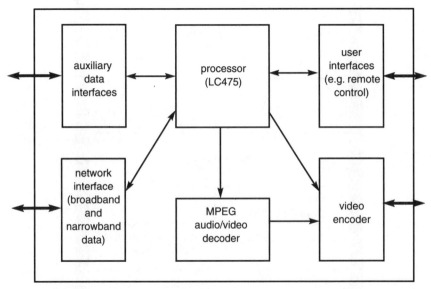

Fig. 6.4   Functional hardware blocks within an Interactive TV set-top box.

An illustration of the layered **software architecture** within an STB is shown in Fig. 6.5.

The software within the Interactive TV set-top boxes adopted a strictly layered approach. All custom hardware within the STB is configured, and supported by custom software device drivers (system extensions in Apple's MacOS™ terminology). All devices are thus made available through the operating system (in this case a reduced version of Apple's MacOS Operating System) to the application layers.

The device driver, and operating system, software is held as resident software within the STB (in either flash or standard ROMs). Other software is downloaded during a boot process (see section 6.2.3.3 below for further details) by a small resident bootstrap application.

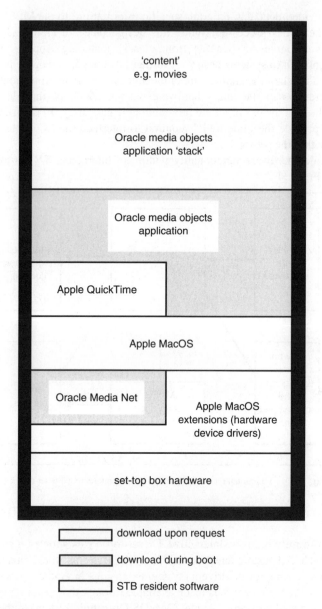

Fig. 6.5   Layered software architecture of an Interactive TV set-top box.

The application environment for the Interactive TV STBs is provided through the use of Oracle Media Objects™ (OMO). OMO is an object-oriented 4th-generation language (4GL) specifically designed to handle both stream- and file-based multimedia information (e.g. an MPEG encoded video stream can be represented as an object within Oracle Media Objects) within a client server environment. Applications within OMO exist as stacks of cards in much the same way as Hypercard™ applications (a stack is a collection of cards, a card being a screen object consisting of both multimedia presentation objects (pictures, sound, buttons) and script code (intelligence for controlling a user's interaction with the card)).

The synchronization and display issues of multimedia information are handled by Apple's MacOS Quicktime™ and QuickDraw™ extensions.

The network transport functionality within an Interactive TV STB is provided by Oracle Media Net™ (OMN). OMN supports client/server transactions including the download of both applications (OMO stacks) and also MPEG streams.

MPEG video streams decoded by the STB, are displayed upon the back plane of two display planes. The front display plane is available via the operating system for the display of graphics, with an additional feature of setting any pixel to be transparent (i.e. allowing the video plane 'behind' to be seen). This concept is shown in Fig. 6.6. The operation of the two display planes is under software control, using the inherent capability of MacOS to support multiple displays.

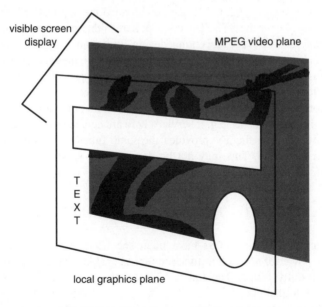

Fig. 6.6   Set-top box screen display planes.

In general, the environment offered to applications within an STB can at best be described as hostile. This is because the STB application must provide excellent response times while performing extremely complicated multimedia operations upon a system with 4 Mbytes of RAM (this memory must fulfil both application and operating system requirements). The design of the STB allows for the RAM to be expanded, but due to the high cost of memory devices this option was not taken.

6.2.3.3   STB functionality

The Interactive TV set-top boxes exist in a number of operating states, with the functionality of the STB determined by the current state. Figure 6.7 shows the major operating states in an Interactive TV set-top box. The initial ('boot') states are required to successfully download and execute an application. The functionality of the STB during application execution is dependent upon the application (notwithstanding the inherent capabilities of the STB). During STB 'boot', the user is not required to perform any operations, the concept being the same as for other domestic equipment (compare a domestic video cassette recorder). The STB will, however, accept a 'stand-by' instruction from the user.

While the STB is in fact a computer, the ethos used is to shield the user from this fact wherever possible.

The early stages of STB operation are to prepare the device for application download, as shown in Fig. 6.7. Once downloaded the application has (almost) complete control over the STB. The concept here is that of painting upon a clean piece of canvas. The STB (accepting design limitations) is thus whatever the application requires it to be. This provides advantages to the service provider, but potential disadvantages to the user, i.e. each service provider may implement their own 'look and feel', or 'branding', of the user interface. This flexibility is a potential confusion for the user, and does not exist with other domestic CPE (customer premises equipment), e.g. if services here are equated to channels on a TV, the remote control on the TV provides the same functions independent of the channel being viewed. This may not be the same for services obtained via an STB.

6.3   FUTURE SET-TOP BOXES

The Interactive TV trial STBs have been specified to ensure that they are sufficiently powerful and rich in functionality to enable a wide range of services to be successfully tested. The future STB for the mass market must be cost effective and, to that end, any redundant or expensive features and functionality would be removed to ensure the entry cost to the consumer is kept to a minimum.

The specification for future STBs must comply with global standards to ensure large production volumes and hence enable prices to fall quickly. This will allow the customer to select the product of their choice from any high street consumer/electronics retailer.

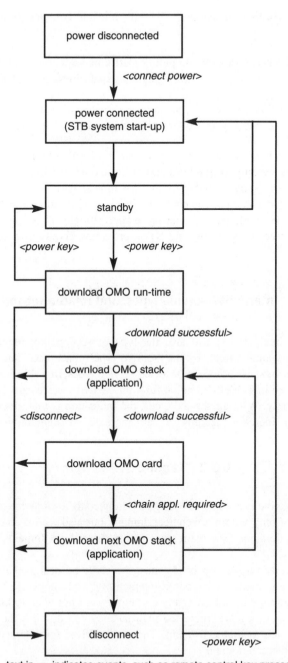

text in <> indicates events, such as remote control key presses

Fig. 6.7   Major operating states for an IMS set-top box.

It is expected that residential STBs will split into two categories in the near term:

- high-end STBs based on computing platforms with considerable expansion capabilities and high-resolution graphics capabilities, i.e. high price;

- low-end STBs for 'VoD-type' simple services with minimal 'application download' capability and low-resolution graphics, i.e. low price.

It should be borne in mind, however, that at the time of writing, digital interactive services are only available as part of limited trials. Limited interactivity is due to be commercially available for the first time as part of either analogue services (interactive TV, analogue cable services), or as an add-on to 'digital satellite' delivery (to enable pay-per-view type applications). These market areas are generally focused towards low-end STBs providing access to a single service (provider).

6.3.1   Interactive service types, and related set-top boxes

The level of interactivity, and also the type of set-top box required, can be categorized, to some extent, by the 'service type' concerned. The relationship between these areas is shown, for selected service types, in Table 6.2.

As indicated in Table 6.2, digital fully interactive services are currently limited to trials only. With all service types, the indication is that, over time, functionality will increase, leading to an increase in interactivity and STB complexity.

6.3.2   Future STB drivers

Future 'interactive' set-top box design is subject to a wide range of driving influences, not only within a technical domain, but also within commercial and consumer-led domains. The main areas within these three domains are given in Table 6.3.

Future interactive set-top boxes must therefore be designed with much in mind. The requirement for increased interactivity, via a network, is not a main consumer requirement. Indeed, many consumers will not want to know how the system works — they are interested in what they can do with the system, i.e. what services are available to them. The major driver for interactivity is thus from the commercial domain.

Table 6.2   Table of service types and set-top box interactivity.

Service type	Status of service	Interactivity	Set-top box required
'Interactive TV'	Service started in the UK in late 1995.	Download to STB via broadcast. Return channel via PSTN modem. Interactivity is for service selection, and feedback of user input (also 'scores' for interactive game shows).	Low-end (current trials are initially using high-end STBs based upon a PC platform)
Video on Demand	Trials only	Download/return channel provided by broadband feed to STB. Interactivity is for programme selection, including 'virtual VCR' controls.	High-end (may in future be pre-configured low-end STB).
Services on Demand	Trials only	Download/return channel provided by broadband feed to STB. Interactivity is dependent upon the functionality of the service (application) being used.	High-end
Digital Satellite	European service now operational (Canal+, a French broadcaster)	Limited download to STB via broadcast, return channel via modem. Interactivity is (initially) for customer authentication, and for pay-per-view. Later will support: text/still pictures on demand; transactions; messaging.	Initially low-end, single-purpose STB. May require more complex STB if greater interactivity is used in later services.
Digital terrestrial TV	UK service due to start in early 1998	Limited download to STB via broadcast, return channel via modem. Interactivity is (initially) for customer authentication, and for pay-per-view. Later will support: text/still pictures on demand; transactions; messaging.	Very likely to be included in digital satellite STB.
'Interactive Cable'	Digital — trials only (e.g. Westminister TV) Analogue — services currently in operation in a number of locations (e.g. Videotron in London)	Download to STB via broadcast, limited cable return channel (TDM QPSK). Interactivity is for service selection (including software download), and feedback of user input.	Initially high-end (based upon IBM PC), although cost reduction will follow.

Table 6.3   Table of drivers for future set-top boxes.

Technical Domain	Commercial Domain	Consumer Domain
Effect of regional and international standardization	Need for high sales figure	Need for low purchase/ser-vice price
Disparity of network transport architectures	Manufacturer profit margins, on a high cost (to produce) item	Need for ease of use
Introduction of other digital services leading to convergence of CPE	Product differen-tiation through additional 'features'	Need for wide range of ser-vices
	Cost of service roll-out	Adverse reaction to computer technology in the home
	Network bandwidth cost for broadband service	
	Determination of real consumer demand	
	Service provider differentia-tion through 'branding'	
	Use of interactivity to enable customer billing (e.g. pay-per-view)	

6.3.2.1   Commercial realities

While a proportion of consumers will invest in new technology (the 'early adopters'), the majority will invest in a mature market (an example of this can be seen in domestic video cassette recorders, and also UK satellite decoders). Cost models have predicted that consumer interest will reach 'critical mass' when the retail price for an entry-level, low-end interactive set-top box reaches £200 (or equivalent). This 'critical mass' price is based upon a number of commercial models produced, and also from observation of the UK satellite receiver market (here the £200 threshold was inclusive of installation).

Meeting this price is technically possible, although a great deal of cost reduction must be achieved.

Current cost estimates for a low-end STB would indicate that some of the critical components (microprocessor, MPEG decoder) will need to be reduced to 30-50% of their current cost in order to achieve the required target. The cost reduction required for a higher functionality (high-end) STB is, however, not so critical.

It is essential to 'seed' the market in early stages (as with other similar services) to ensure a 'virtuous circle' — whereby the number of STBs sold encourages further service providers to become involved, who in turn encourage further customers to buy STBs to receive the services. This is depicted in Fig. 6.8.

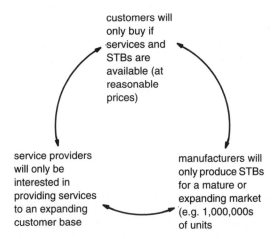

Fig. 6.8   The 'virtuous circle'.

This circle must be initiated through the adoption of risk. In other comparable services, the service provider adopts the risk for the production of equipment for supply to the consumer. In this instance, the service provider is considered to act as a 'broker' for a number of content and information providers. The adoption of risk by the other two parties, for instance 'early adopter' consumers and 'innovative' suppliers, is unlikely to 'kick start' a market in this area.

The average consumer is distrustful of products that 'look' like computers. Many computer system manufacturers interested in moving into the future 'intelligent interactive set-top box' market, view such devices as Trojan horses, i.e. a means of selling a home computer to consumers who would not normally purchase one. Such devices would be capable of upgrade at a later date to offer full computer facilities. Apple's Pippin™ games console is one example of a commercially available device with such a capability.

6.3.2.2   STB convergence

There is continuing progression towards digital services in the home. Currently games and CD-ROM are delivered, in digital format, from local mass storage. It is only a matter of time before satellite, terrestrial and cable broadcasters switch to digital transmission in order to make more efficient use of the spectrum.

The customers for digital video services will not want to buy an STB on a per application basis — the stack of consumer electronics in the average home is already in danger of 'toppling' over. For this reason it is important that the popular services are rapidly identified and that STBs offering the most common configurations are available. In the short term, it is expected that some early digital

STBs will have an expansion socket to allow for services to be added in the future via an adapter module.

As soon as the dominant combination of digital services is known there will be an integrated STB for the dominant package of services. When broadcast digital terrestrial TV is commonplace, the STB will cease to be a separate unit as all of the functionality will be absorbed into the TV. The likely combination in the future is to combine the following services into the TV:

- digital broadcast (probably terrestrial);

- Internet access;

- 'superhighway' socket for highly interactive and 'on-demand' services;

- local high-density storage such as CD-ROM for rarely updated information.

Games consoles will have an integrated 'superhighway' socket for customers whose first requirement will be for game playing, either in stand-alone mode, or more likely competing with other players via the network. STBs will have the ability of downloading games applications from the server to a games console for remote game playing.

Further steps towards convergence are evident with the introduction of the network computer concept. Network computers (NCs) provide a series of interactive informational services (e.g. Internet browsing and e-mail) via a low-cost STB, and displayed on a domestic TV. Service delivery to the NC STB is via a PSTN modem.

6.3.2.3   Content providers

The importance of the content provider cannot be overstated. Without sufficient content availability, the consumers will not buy an STB and without a potentially large customer base, content will not be made available.

One major concern of the content providers currently, relates to the high quality of the digital material, which makes piracy of the content attractive. It is important to identify 'copy protection' procedures within the STB to prevent digital copies of high-value content being taken. There are several methods of accomplishing copy protection through scrambling and conditional access or by 'fingerprinting' recorded content such that the source is known and it would be possible to only allow replay through STBs at the same address.

6.3.3   Standardization

It is essential to establish world-wide standards for interactive STBs to enable the manufacturers to invest, with confidence, in the necessary enabling technology to ensure that STB prices fall quickly. To this end BT are currently supporting the consensus-forming and standards bodies listed in Table 6.4.

Table 6.4   Table of bodies relating to set-top box standards.

Standards body	Name	Area of interest
Digital Audio-Visual Council	DAVIC	Global multimedia systems specifications
Digital Video Broadcast	DVB	European satellite and cable specifications
Asynchronous Transfer Mode Forum, Residential Broadband Group	ATM Forum, RBB	Mainly US body, ATM residential network transport and interfaces
Video Electronics Standards Association (VESA) open set top	VESA-VOST	American standards body, currently looking at physical interfaces and wiring, e.g. home LANs
Moving Picture Experts Group	MPEG	Digital coding of moving pictures
Digital Storage Media Command and Control	DSM-CC [3]	User/network and user/user control (working group of MPEG)
Multimedia and Hypermedia Experts Group	MHEG [4]	Coding and interaction of multimedia objects

Standardization is required to ensure interoperability of servers and STBs from a range of manufacturers, independent of the access network. There are currently two schools of thought with regard to the interoperability of STBs within the overall system.

- Standardization of the format in which objects are delivered to STBs — in this way STB manufacturers know exactly what they have to deal with. The down side is that the STB must become more interpretative by nature and hence the speed of presenting data may be impeded. This problem will recede as STB processor performance increases for the same cost and as the designers become more experienced.

- The server downloads an application and extensions according to the STB make and model, which is passed to the server from the STB during initialization. Thus there are islands of interoperability, where it is likely that some STBs will not operate on some servers if the service provider does not choose to support those STBs.

6.3.4   Future STB technology

6.3.4.1   Architecture — hardware and software

Hardware architecture — the hardware architecture of a future STB is unlikely to vary much from the generic model shown in Fig. 6.4. The major area for change here is in the integration of functionality to achieve cost reduction. Already many STB manufacturers are considering the possibility of highly integrated STBs consisting of a minimal chipset. A three-chip solution is considered feasible by many manufacturers.

The hardware architecture chosen for a future STB will ultimately depend upon the 'model' chosen. Currently two extreme models exist — the hardware processing and the software processing models. The hardware processing model assigns hardware blocks to perform functional tasks (e.g. MPEG decoding being performed by a dedicated MPEG decoder chip). The software processing model assumes one or more high-speed microprocessors providing equivalent functionality. The latter approach is more flexible, but is also arguably more expensive.

One requirement for a future STB will be the ability to connect to a number of differing service provision media (see section 6.3.2.2), if only because the core of a particular unit may be manufactured for use in a number of countries. The concept of a modular STB with two main modules (the first module comprising the network interface, and the second module comprising all other functions of the STB) is thus likely to be adopted for all but the most basic STBs in the future. The STB will have provision for a conditional access module. This will either be integrated within the unit or, more likely, be a replaceable module, probably via the DVB common interface. The network interface module would enable the remaining STB functionality to be network-independent.

The existence of a number of 'grades' of future STB is a certainty. Low-end STBs will be heavily integrated, with minimal functionality and little or no ability to upgrade. Such STBs are likely to be based upon the 'three-chip' architecture. High-end STBs, however, are more likely to be based around a software processing model, with the capacity for expansion.

Software architecture — the move for greater conformance between interactive multimedia service platforms (see section 6.3.3) will require some form of compliant application interface to be defined. While it is possible to transcode an application into a number of formats (the correct format for an STB being downloaded as required), the appeal of a portable application format has been recognized. This form of coding application would be independent of the application-creation tools and language, and would also hide the STB operating system and hardware. A number of such coding formats exist, or are being developed (e.g. MHEG Part III Script Interchange Format and Java™).

The grading of STBs is expected to heavily affect the software architecture; low-end STBs are most likely to be offered supporting only a limited number of services (in that the low-end STB will not support full application download). The layered model shown in Fig. 6.9, would thus be compacted, with the 'middleware' layer absent.

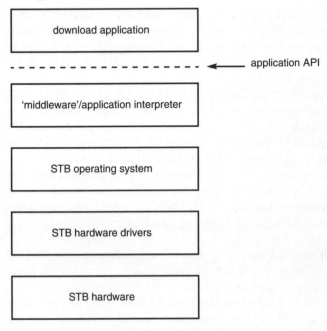

Fig. 6.9   Future set-top box software architecture.

6.3.4.2   Network interface

The STB of the future is likely to be connected to an ATM network and, as such, it is expected that the ATM Forum will take the leading role in specifying a low-cost physical interface for residential STB applications. The STB network interface will be in the region of 25 Mbit/s symmetrical, although the highest bandwidth entertainment application will probably be high-definition television (HDTV) which will require around 12 Mbit/s.

The STBs developed for the Interactive TV trials have no call set-up capabilities, but rely on 'nailed up' virtual circuits (VCs). The STB is then operated in 'user-to-user' mode to control 'sessions' and navigate through applications.

The emerging DAVIC STB standard is expected to favour DSM-CC user-to-network signalling for call set-up, with translation to the appropriate B-ISDN or ATM protocol stacks being accomplished within the network. The ISDN E.164

specification and MAC addresses are expected to be adopted for addressing STBs.

6.3.4.3   MPEG decoders

The MPEG1 standard is used for most stored content such as CDi and CD-ROM. The total bandwidth is 1.5 Mbit/s-1.2 Mbit/s for video and 256 kbit/s for audio. MPEG2 was developed mainly for real-time encoders for broadcast systems and if main profile/main level is used, as specified by DVB, it is backward compatible with MPEG1.

The majority of the early digital STBs (CDi and CD-ROM players) utilize MPEG1 decompression, based on a hardware chipset normally comprising a demux IC to demultiplex the audio, video and data streams, an MPEG1 video decoder and an MPEG1 audio decoder. There are already single-chip solutions for this functionality both for MPEG1 and MPEG2 decoders.

MPEG2 decoders are used in digital broadcast receivers both in Europe and North America, although there are also proprietary solutions in use such as General Instrument's Digicypher™. MPEG2 encoders/decoders are able to dynamically change bit rates to enable the most appropriate bandwidth to be used according to the source content requirements, i.e. talking heads require much less bandwidth than fast-moving sports programs such as tennis. MPEG2 decoders are also able to decode MPEG1 content thus ensuring that the large library of content being generated for local storage systems is still usable.

In the future, the decoding algorithms will probably be executed on the main CPU together with some hardware-assist circuitry to promote a higher degree of product integration.

6.3.4.4   Physical user interfaces

Future STBs are likely to provide a range of interface options. The basic, or entry level, physical interface will remain the infra-red remote control. The physical user interfaces available, however, are likely to be legion, depending mainly upon the users' requirements, and the constraints of the overall service interface. A potential list of other interface devices could include:

- keypad/keyboard;
- games;
- pads;
- voice activation;
- air mouse;
- personal digital assistant/notepad computers.

Other devices are also likely to be attached to the set-top box for data entry purposes; once again the list of such devices is legion and may include:

- printer;

- PC;

- CD-ROM drive;

- disk drive;

- games console adapters;

- encoder/camcorder input;

- digital video cassette recorder.

Service offerings, and consumer requirements will be the main determining factors in what may be a truly diverse market.

6.3.4.5   Graphical user interfaces

STB graphical interfaces will vary depending upon the service and complexity of the STB. Low-end STBs are likely to have a limited graphics plane (e.g. 16 colours, 320×240 pixels) which will determine the format of the graphical interface chosen.

High-end STBs are likely to offer 'workstation' quality graphics (e.g. 32767 colours, 640×480 pixels minimum), enabling more complicated interfaces to be supported. This will provide the ability to generate a number of differing styles of interface, allowing the user to choose the style of the user interface according to their personal preferences (e.g. 'cartoon'-type interface for children). This assumes that the application downloaded to the STB can support a choice of user interfaces.

The option of combining functionality in single integrated circuits may be a deciding factor here. Already some MPEG decoder ICs provide for a simple graphics overlay, thus eliminating the need for separate video graphics memory. Combined devices such as these are likely to offer CCIR601 resolutions (or sub-multiples), e.g. 720×576, 360×288.

6.4   HOME WIRING

The STB requires to be connected to the network, in a manner acceptable to both customer and access network provider. As digital STBs become more widely used in the home environment, customers are likely to install further STBs or home computers. The sharing of resources (e.g. applications and digitally encoded content) between these devices will require a standardized manner of interconnection. To achieve interconnection between STBs, the consumer's home must be wired to provide a simple local area network.

There is considerable debate on the best approach to home-wiring systems. The simple answer is that there is no single approach that will meet everyone's needs and there is also a considerable degree of personal choice in deciding which system to use. It is intended to specify a simple, low-cost home-wiring solution which is easily installed on a DIY basis, causing minimal damage to existing decor. To this end the choice of cable is Category 5 unscreened twisted pair (UTP), because it has a small diameter and can be formed around quite tight corners, and it is relatively easy to terminate with simple tools. Category 5 UTP can support a 155 Mbit/s bandwidth, and can economically support bandwidths in the range of 25-51 Mbit/s.

A symmetrical system is preferred allowing broadband content to be transmitted in both directions, this will allow for the addition of videotelephony in the future as the costs of this CPE become attractive to the customers.

It is expected that the home network of the future will be based on an asynchronous transfer mode transport layer, thus enabling individual pieces of CPE to be uniquely addressed across the network.

6.4.1   Entry-level system

An entry-level home-wiring system will simply consist of a point-to-point connection between the network terminating equipment (NTE) and the STB. The STB should then have a high-speed data port, e.g. IEEE 1394, to enable download of data and programme information to a second terminal, such as a games console or PC.

6.4.2   Mid-range system

A typical mid-range home system will probably comprise two STBs (one of which may be a PC) and a digital video cassette recorder (DVC), thus enabling the consumer to have a one-to-many platform in the lounge via the TV, and one-to-one platform elsewhere in the home via a second TV or a PC. The one-to-one platform would be used for transactional and personal services such as banking, e-mail and training, etc, while the one-to-many would be used for reception of broadcast or on-demand broadband video content. Both platforms would be able to record and play back material via a common DVC.

6.4.3 High-end system

The ultimate home network will comprise a switched hub into which all the mid to broadband communications CPE within the home would connect. This would enable communications between similar CPE within the home to be established, in addition to routeing incoming calls to the required piece of CPE. The cost of this equipment must be considerably lower than an STB, probably about 50% of the cost of a low-end STB, i.e. around £100. It is possible that this functionality may be incorporated into a PC card decoder at a small additional cost.

6.5 CONCLUSIONS

While existing interactive STBs are bespoke items, designed for specific trials, future STBs will be standardized commodity items. Although the evolutionary path between the present and future STB designs may be clear technically, it will, however, be subject to a number of commercial pressures.

The first generations of interactive STBs are likely to have a retail price from around £400 (DVB Broadcast Receiver) to £500 (STB with application download capability). It is thought that this must be reduced to a retail price of £200 to £300.

It is expected that the most likely STB configuration, in the next five to ten years, will integrate broadcast digital terrestrial or digital satellite services with interactive 'superhighway' services (at varying bandwidths). These are likely to be combined with local high-density storage media, such as CD-ROM or DVDs, for static and 'favourite' information. The integration of these services will enable future content and service providers to produce truly innovative services that will excite and stimulate all sectors of the population.

APPENDIX

List of Acronyms

ADSL	asynchronous digital subscriber line
APON	ATM passive optical network
ATM	asynchronous transfer mode
B-ISDN	broadband integrated services digital network
CDi	compact disk interactive
CD-ROM	compact disk — read only memory
CPE	customer premises equipment
CPU	central processor unit

DAVIC	Digital Audio-Visual Council
DSM-CC	Digital Storage Media — Command and Control
DVB	digital video broadcasting
DVC	digital video cassette
DVD	digital video disc
FPGA	field-programmable gate array
HDTV	high-definition television
IC	integrated circuit
ISDN	integrated services digital network
MAC	media access control
MHEG	Multimedia Hypermedia Experts Group
MPEG	Moving Picture Experts Group
PAL	phase alternating line
QPSK	quadrature phase shift keying
RAM	random access memory
ROM	read only memory
SCART	common name for peritelevision connector
SCSI	small computer systems interface
STB	set-top box
TDM	time division multiplex
TV	television
UHF	ultra high frequency
UTP	unscreened twisted pair
VCR	video cassette recorder
VESA-VOST	Video Electronics Standards Association — VESA Open Set Top
VoD	video on demand

REFERENCES

1.   Moving Picture Experts Group: ISO/IEC11172 — MPEG1 and ISO/IEC13818 — MPEG2.

2.   Cole N G: 'Asymmetric digital subscriber line technology — a basic overview', BT Technol J, _12_ , No 1, pp 118-125 (January 1994).

3.   Digital Storage Media — Command and Control: ISO/IEC13818-6.

4.   Multimedia Hypermedia Experts Group: ISO/IEC13522.

7

BROADBAND MULTIMEDIA DELIVERY OVER COPPER

G Young, K T Foster and J W Cook

7.1   INTRODUCTION

Telecommunications operators' fixed networks were originally constructed entirely from metallic transmission media (predominantly copper), carrying voice-band signals in the region 300 Hz to 3.4 kHz. In developed countries, the core or backbone of the network that interconnects the switching centres is now mainly optical fibre. However, the access portion of the network that connects switches to customers (the so called 'last mile') is still dominated by twisted copper pairs. There are over 560 million twisted copper-pair connections world-wide. The UK alone has around 27 million copper access connections with a replacement cost of over £11 billion. Although optical access-network technologies have been developed and proven, the sheer inertia of the installed copper base means that it could take many years to migrate the access networks from copper to fibre. This chapter describes the transmission technologies available to telecommunications companies (telcos) to deliver broadband services, such as TV, video on demand (VoD) or fast Internet access, over the existing copper access network as it evolves towards a fibre future.

This chapter was previously published as an article in the IEE Electronics and Communications Journal in February 1996.

7.2   THE COPPER ACCESS NETWORK TRANSMISSION ENVIRONMENT

7.2.1   Access network topology

Figure 7.1 shows an overview of BT's access network. The main network consists of large multi-pair cables which radiate out from the main distribution frame (MDF) located within the serving local exchange to flexibility points known as primary crossconnection points (PCPs). Individual cable segments are joined together to form the link from the MDF to PCP. From the PCP outward, the network is known as the distribution network (D-side), whereas the MDF to PCP link is known as the exchange (E-side) network. From the PCP, connections radiate out to distribution points (DPs), often passing through secondary crossconnection points (SCPs) for increased flexibility. From the DP the connection is made to the customer's premises via the dropwire or final drop. It should be noted that the distribution network may be overhead or underground. Underground feed is more reliable and is usually used in modern urban areas. However, in rural areas the high civil engineering costs associated with underground cabling prohibit the use of underground feeds and overhead feeds are more common. Overhead feeds are also used in older conurbations.

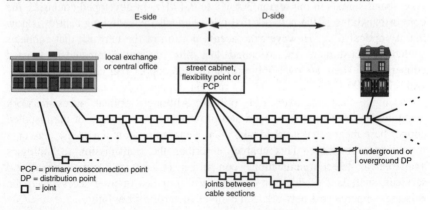

Fig. 7.1   Access network topology.

At the customer's premises there is a demarcation point known as the network termination point or NTP. This is the point at which BT's network officially ends and is a point at which service is provided to the customer via a standardized user-network interface (UNI). For a customer receiving telephony service (POTS) this point is the familiar white master socket.

The cables that are predominantly used in the access network are unshielded twisted pairs (UTPs) of differing gauges. The smaller gauges are found closer to the exchange (to make the large multi-pair cables easier to handle) and the larger gauges are found towards the customer's premises where the environment is more hostile and there is a need to achieve maximum range for a given transmission and signalling-resistance requirement. Copper cables dominate, but aluminium cables are also found.

Typical gauges of cables (from MDF to DP) are : 0.32, 0.4, 0.5, 0.6, 0.63, 0.7, 0.8, and 0.9 mm.

7.2.2   Advances in transmission-channel modelling

The metallic twisted-wire pair was originally specified and designed to carry analogue telephony signals within a 300 Hz to 3.4 kHz frequency band. As frequency rises so does the signal attenuation due primarily to the skin effect which manifests itself as an increasing resistive component per unit length. Most of BT's access cables are made from solid copper conductors sheathed with polyethylene (PE) dielectric. This forms a low but significant capacitance per unit length. For most polyethylene cables, capacitance per unit length does not vary appreciably with frequency and can be considered constant.

It is useful to be able to model the electrical behaviour of the access channel when designing and estimating performance of various transmission technologies. Recent work at BT Laboratories has led to the development of an empirical model for homogeneous access channels. The model has two forms. Firstly, a 7-parameter form covering the region from DC to 2.5 MHz (i.e. asymmetric digital subscriber line (ADSL) operation). Secondly, an enhanced 13-parameter version covering the frequency range DC to 30 MHz (for high-rate studies such as very high-rate digital subscriber line (VDSL)). The 13-parameter model has been validated from DC to 30 MHz with excellent accuracy (typically 0.1 - 0.5 %).

Figure 7.2 shows a plot of the typical insertion loss of 300 m sections of various dropwires types used by BT in its access network. It can be seen that there is a large variation in insertion loss at the higher frequencies (e.g. >30 dB at 20 MHz).

7.2.3   Impairments

Without noise (or impairments) even a bandlimited access channel has infinite information-carrying capacity. In reality there are many sources of noise which collectively will determine the information-carrying capacity of the access channel. In a well designed digital transmission system, the effects of internal system noise will be small. In this case the information-carrying capacity will be determined by the access channel and external noise sources, namely:

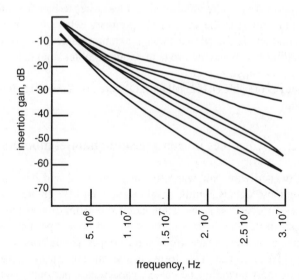

Fig. 7.2   Insertion loss of 300 m lengths of various dropwires.

- near-end crosstalk (NEXT);

- far-end crosstalk (FEXT);

- radio frequency interference (RFI);

- impulsive noise.

Each of the above noise sources will, to a greater or lesser extent (in combination), determine the limits of the information-carrying capacity of the access channel. Systems which are required to deliver very low bit-error rates (BERs) must be designed to operate in a noise environment dominated by these impairments (as well as the channel attenuation itself).

Internal noise sources should not be ignored but they will largely be:

- a feature of the state of developments in signal processing technology;

- an ever-present feature of the passive and active electrical components (e.g. white noise) — designers will never be able to completely ignore the effects of white noise.

7.2.3.1   White noise

Generally, white (thermal, shot, quantization) noise sources are relatively low level and are usually insignificant when compared with other noise sources (e.g. NEXT and impulse noise).

7.2.3.2   Near-end crosstalk

NEXT arises due to signals from transmitters on other pairs operating in the same multi-pair cable. These are signals which interfere with the input of a transceiver at the same end. The transmitted signal leaks into the receiver via capacitive and inductive coupling paths (see Fig. 7.3). NEXT is usually the most significant noise source which limits the reach of high-rate duplex transmission systems.

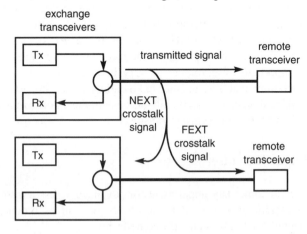

Fig. 7.3   NEXT and FEXT noise coupling.

7.2.3.3   Far-end crosstalk

FEXT occurs when signals from transmitters on other pairs in the same cable leak into the input of the wrong transceiver at the other end (see Fig. 7.3).

FEXT is not usually significant for symmetrical digital transmission systems because the FEXT always suffers more attenuation than the signal passing through the channel. However, it is significant for ADSL systems where the different directions of transmission occupy different frequency bands.

As for NEXT, FEXT is usually worst when the interfering signals on the other pairs are of the same kind as the system being interfered with. This is known as self-FEXT.

7.2.3.4   Radio frequency interference

RFI is a source of noise which affects almost every access wire-pair. It varies considerably from pair to pair and has characteristics which show temporal variations. Digital transmission systems must tolerate certain levels of RFI as prescribed by current European legislation on electromagnetic compatibility (EMC).

Varying amounts of RFI will undoubtedly exist in the same frequency range as the transmitted signal. This interference is in-band and therefore cannot be filtered out. Metallic access-transmission systems should not interfere with radio transmissions. This places a limit on the transmitted-power spectral density. It is fortunate that access-network pairs are well balanced and do not pick up or radiate RFI easily. The longitudinal balance (common mode to differential coupling loss) of most wire-pairs is extremely large at low frequencies (> 60-70 dB) decreasing to around 30 dB at 30 MHz. This means that RF signals will suffer quite large attenuation before coupling into the receiver of a transmission system.

RFI is an important noise source for very high-rate transmission systems which occupy the same bands as amateur radio enthusiasts (1.8 to 30 MHz). This type of RFI is difficult to cancel adaptively because of the on/off nature of Morse and single-sideband suppressed carrier transmission.

7.2.3.5   Impulsive noise

Impulsive noise is another key impairment for advanced access transmission on metallic twisted pairs. It is caused by a variety of sources producing short electrical transients. Such transients may come from:

- household appliances switching on and off (e.g. fluorescent lights, refrigerators, cookers, etc);

- telephony itself (e.g. on-hook/off-hook and ringing);

- arc welders, electric cattle fences or locomotive traction systems.

These disturbances can be electromagnetically coupled into the access network and may cause error bursts in digital transmission.

Forward error correction (FEC) and data interleaving are normally used in ADSL and VDSL systems to mitigate the effects of impulse noise. However, there is a trade-off — increasing protection against impulse noise is usually accompanied by increased transmission delay or latency.

7.2.4   Theoretical capacity

The theoretical information-carrying capacity (in the presence of Gaussian noise) of an access channel may be calculated using the well-known Hartley-Shannon theorem:

$$C = B \cdot \log_2 \cdot (1 + \frac{S}{N}) \text{ bit/s}$$

where B is the bandwidth, and S/N is the mean-square signal-to-noise ratio (SNR) at the receiving station in the channel.

Assuming a base-band signal, then the information-carrying capacity is given by:

$$C = \int_0^{f_{max}} \cdot \log_2 \cdot \left(1 + \frac{S}{N}\right) \mathrm{d}F \text{ bit/s}$$

Using the empirical channel model and knowledge of the behaviour of the more predictable noise sources (e.g. crosstalk), the theoretical maximum information capacity may be calculated for arbitrary channels and noise environments.

With respect to Gaussian noise, both ADSL and VDSL systems are primarily FEXT-limited because they may employ FDM techniques to separate the two directions of transmission (thus removing the unwanted effects of NEXT) — 1% worst-case FEXT is easy to predict given knowledge of the power spectral densities of the interferers and the channel length.

A study of the information capacity of various metallic cable types is quite revealing. For example, Fig. 7.4 shows the FEXT-limited capacity of 0.5 mm D-side underground cabling when appropriate assumptions are made for a VDSL transmission environment. Capacity is plotted as a function of the maximum upper frequency (of a baseband transmission) and channel length. Contours of fixed capacities of 100, 50 and 25 Mbit/s are also shown. The potential for very high-rate transmission over relatively short distance is clearly seen. It must be noted that this theoretical limit will only be approached with transmission systems of infinite complexity, and presumably infinite cost. Practical ADSL systems are achieving typically 30-60% of theoretical capacity predicted by knowledge of the channel and noise environment. VDSL systems are expected to realize similar levels of practical capacity.

7.3   OVERVIEW OF TRANSMISSION SYSTEMS

7.3.1   The ADSL concept

ADSL is a new method of transmitting digital data at high bit rates over the existing installed twisted-pairs of the access network. This technology will release the previously untapped capacity in the copper telephone network as described in the previous sections. When the transmission capability of ADSL is

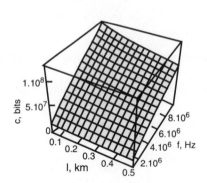 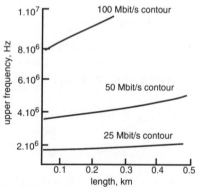

Fig. 7.4   Theoretical capacity of 0.5 mm D-side cabling.

combined with the latest generation of video compression technology, delivery of broadband interactive multimedia services over the copper access network becomes possible.

Voiceband modems are now able to transport 28 to 33 kbit/s through a 4 kHz voice channel. Use of similar advanced modulation and signal processing techniques have given ADSL technology the power to deliver downstream (exchange-to-customer) transmission channels of up to 6.144 to 8 Mbit/s. This is achieved without disturbing the telephony service already installed on the line. A low-speed return (customer-to-exchange) channel enables customers to select and control the downstream information. An ADSL transceiver can be installed simply at each end of a customer's existing copper line.

Marketable interactive residential broadband services are still at an embryonic stage, yet it is already clear that many will be asymmetric in terms of data rate requirement, with a high capacity needed towards the customer and a low capacity in the return direction. Video on demand is attracting much interest because many customers can identify with the concept of having the equivalent of a video rental store combined with a VCR in the network. It would allow customers to select, stop, start, pause, rewind and fast-forward video films. An ADSL access transmission system operating at 2.048 Mbit/s can deliver MPEG- (Motion Picture Experts Group) algorithm digitally encoded video, plus digital stereo sound. Such a system has a subjective picture quality comparable to that of a domestic VCR. A system operating at 6.144 Mbit/s can deliver three VHS quality channels or be dedicated to a single MPEG-encoded session with a subjective picture quality comparable to live broadcast. Other interactive services, such as home shopping, games, education, remote LAN access (teleworking) and high-speed Internet access, could use the same network infrastructure.

The ADSL concept and its application to transport interactive video services was proposed at the beginning of the decade by research workers such as Waring *et al* [1] and Chow and Cioffi [2]. It has since evolved from computer simulations and laboratory prototypes to production systems which are available today and will become increasingly integrated. The principle is simple — transmit a high-capacity channel downstream to the customer and transmit at a much lower bit rate in the return direction from the customer to the network, and do this simultaneously, without interfering with existing telephony on the copper pair.

Since the return transmitter operates at a much lower frequency compared with the forward direction (see Fig. 7.5), the levels of crosstalk at the customer end are low compared to symmetrical data rate systems [3]. The absence of such crosstalk enables a much greater transmission distance to be achieved.

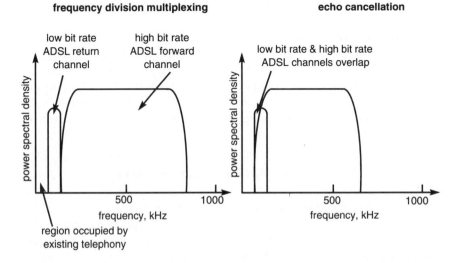

Fig. 7.5   ADSL channel separation.

By employing a passband modulation method an ADSL transceiver can operate at frequencies above those of standard telephony, and then, provided filtering is employed to remove undesirable noise transients such as those due to loop disconnect and ringing, an ADSL system can co-exist with telephony on the same pair. Intelligent signal processing in the ADSL transceiver can adapt automatically to optimize performance on each individual copper line, and track temporal changes, such as those due to temperature, moisture or continuous interference sources. Manual setting up is also avoided.

There exists a family of standardized ADSL systems, with downstream bit rates of 1.536-6.144 Mbit/s. The bit rates for the simplex downstream channels are based on multiple channels with a maximum capacity of 6.144 Mbit/s. This allows up to four possible transport channels of 1.536 Mbit/s or up to three channels of 2.048 Mbit/s (the ISDN primary rate). The ADSL data multiplexing format of the American National Standards Institute (ANSI) ADSL standard [4] has been designed to be flexible enough to allow other channelizations, for example, to transport data in ATM format. Some manufacturers' ADSL implementations will deliver around 8 Mbit/s downstream to accommodate two channels of 4 Mbit/s broadcast video. The transmission reach of a system decreases as the operating bit rate is increased. ADSL systems operating at 2.048 Mbit/s are expected to have around 90% of BT customers within range. Systems operating at the higher rates have a reduced range and thus a lower network penetration.

In addition to the simplex downstream channel, an ADSL system will transport a return or control (C) channel for user-to-network signalling. The standard C channel operates at either 16 kbit/s or 64 kbit/s, although some early systems use 9.6 kbit/s. An embedded operations channel is used to carry network management information.

Two further optional narrowband, duplex channels may be transported simultaneously by standardized ADSL. These can have additional capacities of 160 kbit/s and 384 kbit/s. Some manufacturers' ADSL implementations are capable of providing a 1-Mbit/s duplex channel. Maximum time-delay limits on some services place restrictions on the degree of latency allowed in the ADSL system. For such services, delay is minimized by dispensing with the interleaving of the error-correction code.

7.3.2   The hybrid fibre/copper concept

In time the telco copper networks will begin to include more optical fibre cables, thus ensuring that homes have best possible connectivity for future communications needs. This will be required in the future as interactive multimedia services proliferate. Each member of a family may wish to access such services simultaneously, e.g. one member watching a movie, another shopping from home and children playing games or getting homework assistance over the network.

Researchers are now studying even higher speed access systems for transmission over the shorter D-side copper wires at the customer end of the access network. Work on these high-speed systems is focused on two scenarios for a hybrid fibre/copper access network [2], namely fibre-to-the-cabinet and fibre-to-the-kerb. This technology is being tagged VDSL. This can use the same transmission techniques developed for ADSL. However, due to the shorter lengths of copper involved in only connecting from the customer to a kerbside cabinet, higher transmission rates are achievable.

In the fibre-to-the-kerb architecture, optical fibre is deployed as far as the DP or last joint box in the D-side of the network (refer to Fig. 7.1). This requires the VDSL transmission system to operate over 'copper tail' distances of around 50 m to 300 m (Fig. 7.6). Manufacturers and standards forums are examining data rates in the range 25 Mbit/s to 51 Mbit/s for this scenario. The second architecture being considered for VDSL is a hybrid fibre/copper network where optical fibre is only deployed as far as the cabinet. This requires VDSL transmission over a distance of around 1 km. Data rates up to 25 Mbit/s are being examined for this fibre-to-the-cabinet architecture. Downstream data rates of around 12, 25 and 51 Mbit/s are of interest because of their relationship to transport rates being examined by the ATM Forum. Upstream data rates are mainly focused on the range 1.5 Mbit/s to around 12 Mbit/s. Initial investigations have shown fibre-to-the-cabinet as being the more cost effective.

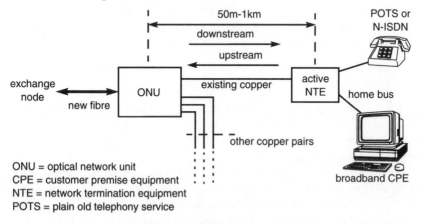

Fig. 7.6   Hybrid fibre/copper access network.

Remotely sited VDSL using an optical-fibre feed, could remain an access network option for some years. This high-speed copper transmission could play a significant role in bringing multiple channels and higher quality channels such as for HDTV to those customers who require them.

7.4   TRANSMISSION TECHNIQUES

7.4.1   Separation of downstream and return channels

Two directions of transmission have to be achieved on the same twisted-pair. Early ADSL systems exploit frequency division multiplexing (FDM) as shown in Fig. 7.5. The low-bit-rate control channel usually occupies the frequency range immediately above the voiceband region of telephony. The high-bit-rate forward or downstream channel occupies the higher frequencies. The total operating frequency range for ADSL is usually limited to about 1 MHz by encoding several bits per symbol. The channels in Fig. 7.5 are shown with flat spectral distribution. In reality this may be varied in order to maximize performance. For VDSL the spectra could theoretically extend as high as 30 MHz, constrained by

EMC considerations. Practical systems will, however, probably not extend much beyond 10 MHz. The VDSL spectra may start at a much higher frequency than ADSL (e.g. nearer to 600 MHz instead of 20 kHz). There are three advantages to this:

- it could avoid the radio band used for AM broadcast transmissions thus avoiding a potential RFI noise source in certain locations;

- it allows the use of simpler (and hence smaller) filters to split the VDSL signals from baseband services such as POTS — this is important since one end of the system would be located in an external street unit;

- it allows both POTS and Basic Rate ISDN systems to exist below the VDSL signals.

This facilitates graceful network evolution by initially using the hybrid fibre/copper network as a broadband overlay. Existing services such as basic rate ISDN could then later be migrated by embedding them within the payload of the hybrid fibre/copper broadband delivery system.

For ADSL there is current interest in allowing the high bit rate forward channel to overlap with the low-speed return channel (Fig. 7.5), in order to take advantage of the reduction in cable loss at the lower frequencies.

This may be achieved using asymmetric echo cancellation which models and cancels the interference between forward and return channels. The ADSL standard [4] allows for interworking between different systems by establishing either the FDM or echo-cancellation method, during the start-up sequence. Asymmetric echo cancellation is particularly new for discrete multi-tone (DMT) ADSL transceivers and many manufacturers are choosing to omit this option in their initial developments. The additional complexity of echo cancellation is expected to yield around 2 dB of performance improvement. The use of echo cancellation effectively moves part of the filtering problem from analogue to digital signal processing which is more amenable to VLSI. The benefits of echo-cancelling ADSL systems are greater when the control-channel bandwidth is expanded to carry a larger payload such as 384 kbit/s videotelephony. In such circumstances an FDM implementation forces the downstream channel to occupy higher frequencies with subsequent increased attenuation and hence reduction in range. It should be noted that overlapping the two ADSL channels by the use of echo cancellation introduces the possibility of an additional noise source known as self-NEXT (see section 7.2.3.2), which does not occur in FDM implementations. This will be more acute for large-scale ADSL deployment where the possibility of other ADSL-crosstalk interference is increased. However, as long as the transmit PSD is controlled to be at a similar level to other systems occupying the overlap frequencies (e.g. HDSL) and the return-channel bandwidth is not in excess of these other systems, then the worst-case multiple-crosstalk models used in many specifications, standards and deployment planning rules will still be valid.

The coexistence of ADSL on the same line as existing telephony presented a major technical challenge. ADSL signals of less than 1 mV need to be recovered in the presence of ringing and loop disconnect pulses of many tens of volts. Hence the design of the separation filter requires very high stopband attenuation. The potential impact on telephony of the ADSL separation filter is increased echo and sidetone [5].

Three transmission methods were originally proposed for use for ADSL and are now all being considered for VDSL. These options are briefly outlined in sections 7.4.2-7.4.4 below.

7.4.2   Quadrature amplitude modulation (QAM)

QAM is a well-established and documented modulation technique used extensively in voiceband modems and microwave radio systems. The source data is split into two half-rate streams which are then modulated on to a pair of orthogonal carriers for transmission.

The orthogonality is provided by a sine and cosine mixing function (modulator) which can be implemented digitally. The transmitter comprises a scrambler, bit-to-symbol encoder, transmit low-pass filter, modulator and digital-to-analogue converter (Fig. 7.7).

At the receiver, the orthogonality allows the two bit streams to be separated by demodulation for subsequent detection of data. Detection involves identifying the received 2-D complex symbol and then mapping this back into binary and descrambling the bit stream. The receiver also includes equalization to compensate for the dispersion caused by the channel. In order to avoid aliasing in the analogue-to-digital conversion process, the front-end of the receiver would typically operate at either 3 or 4 samples per symbol which implies the need for a fractional-tap linear equalizer that operates at 3 to 4 times the symbol rate. Linear equalizers are well known to cause noise amplification in data transmission systems. Hence, most early proponents of QAM ADSL examined the use of decision feedback equalization (DFE).

7.4.3   Carrierless amplitude/phase modulation (CAP)

CAP is a 2-D transmission scheme similar to QAM. CAP has the same type of spectral shaping as QAM and can be made compatible with QAM. However, instead of orthogonal carriers being generated by a sine and cosine mixer, modulation of the two half-rate bit streams is achieved via two digital transversal

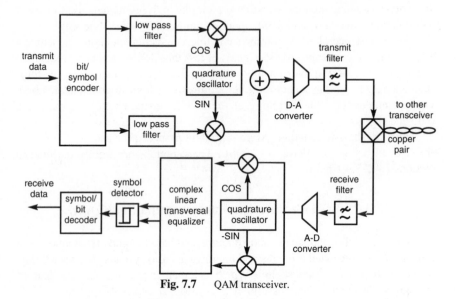

Fig. 7.7   QAM transceiver.

bandpass filters that have impulse responses which are a Hilbert pair (same amplitude characteristics but phase response differs by $\pi/2$). The transmit signal is then formed by combining the two outputs of the two digital filters (Fig.7.8).

The receiver must perform an equalization function and can use the same type of equalizer structures used for QAM, e.g. two parallel linear adaptive filters with fractionally spaced tap coefficients and possibly a complex adaptive feedback filter (DFE). The front-end fractional tap filters in such a receiver also provide the matched filtering function to separate the incoming line signal into the two separate orthogonal components.

7.4.4   Discrete multi-tone modulation (DMT)

DMT is a form of multicarrier modulation. DMT divides time into regular 'symbol periods', each of which will carry a fixed number of bits. The bits are assigned in groups to signalling tones of different frequencies. Hence, in the frequency domain, DMT divides the channel into a large number of sub-channels. The line capacity varies with frequency and those sub-channels with higher capacity are assigned more bits. The bits for each tone or sub-channel are

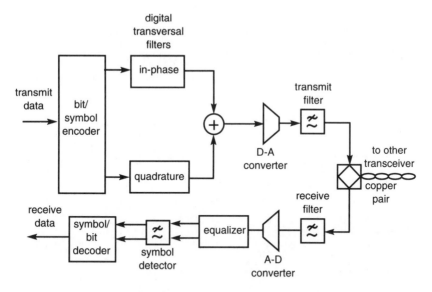

Fig. 7.8   CAP transceiver.

converted into a complex number which will set the tone's amplitude and phase for the symbol period. Hence, conceptually, DMT can be thought of as a bank of contiguous QAM systems operating simultaneously in parallel, each with a carrier frequency corresponding to a DMT sub-channel tone frequency (Fig. 7.9). Thus the DMT transmitter essentially modulates data by forming tone bursts for a number of frequencies, adding them together and sending them to line as a 'DMT symbol'.

The multicarrier modulation (and demodulation) requires orthogonality between the various sub-channels. This can be implemented in an efficient all-digital realization by exploiting fast fourier transform (FFT) methods [6].

Another form of multicarrier modulation being explored by some manufacturers uses a wavelet transform instead of the FFT. This is known as discrete wavelet multitone (DWMT). With both DMT and DWMT the number of bits of data sent over each sub-channel can be varied adaptively depending on the signal and noise levels in each sub-channel. Not only does this allow performance to be maximized for each particular loop but can be a particular advantage for loops which suffer crosstalk or radio-frequency interference, since the system can adapt automatically to avoid regions of frequency spectrum that are unsuitable for data transmission (Fig. 7.10).

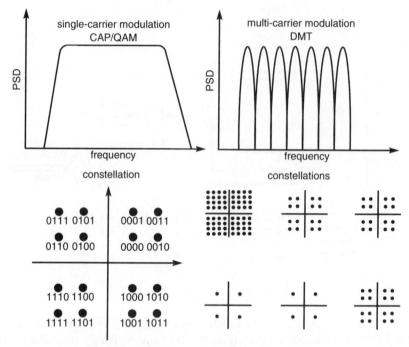

Fig. 7.9   Single-carrier and multi-carrier modulation.

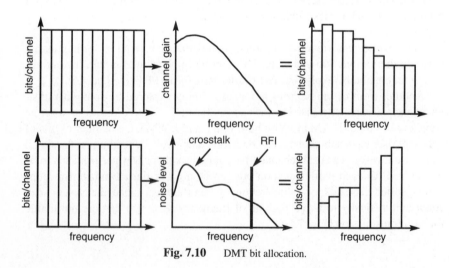

Fig. 7.10   DMT bit allocation.

7.4.5   Improving performance in the real network

Although ADSL systems use a higher absolute transmit power (at +20 dBm) than previous copper pair digital transmission systems, this power is spread over a wider bandwidth. Consequently, the power spectral density (PSD) can be specified to be at similar levels as existing services. The result is that crosstalk between say ADSL and ISDN will be no different from self-crosstalk between two ISDN systems running in the same cable. In fact, the ability of DMT ADSL to allocate transmit power in a non-uniform manner across the sub-channels (within the constraints of upper limit mask) can give even greater spectral compatibility. The system will simply use less power at those frequencies occupied by other services. CAP-based ADSL can use the design of the transmit filter to provide a degree of non-uniform transmit PSD.

ADSL and VDSL transceivers employ many of the state-of-the-art modem techniques that were previously only possible at voice-band signal-processing speeds. As such, ADSL and VDSL are far more sophisticated and adept at coping with the copper-pair transmission environment than currently deployed systems such as ISDN or pair-gain systems.

DMT is a modulation technique that has a degree of inherent immunity to impulse noise. Impulses occurring within a DMT symbol period will often not cause an error. The use of FEC and interleaving gives DMT transceivers resilience against even longer impulsive noise bursts. In designing the parameters of the FEC and interleaving, there is a trade-off between degree of protection and the resulting transmission latency. It should be noted that some services that will be carried over ADSL may have a degree of self-protection against transmission errors. For example, with a video-on-demand service, the video compression scheme may include its own error concealment features.

The main purpose of the FEC and interleaving is to provide error protection against impulsive noise. Laboratory measurements of ADSL and VDSL performance against crosstalk will include the performance enhancement due to the FEC and interleaving (typically around 3 dB). However, in designing deployment guidelines for ADSL it is prudent to determine the system's range in a pessimistic crosstalk environment with the FEC gain 'factored out'. This will ensure that when the system is deployed on the longest of lines, it will still have the power of the FEC in reserve to deal with impulsive noise events, many of which could be sourced from within the customer's premises. If this philosophy is adopted, then the system deployment range can only be increased by enhancing its performance against the steady-state background noise such as crosstalk. One way of achieving this is to use trellis coding. This will give an additional 2.5 dB performance gain when used in conjunction with the FEC.

The front-end of a QAM or CAP transceiver usually contains a fractional-tap equalizer. This oversampling digital filter uses an adaptation algorithm to minimize noise at the decision device in the receiver. In addition to minimizing the intersymbol interference (i.e. performing equalization), this filter can also synthesize a notch in its passband to reduce the impact of narrow-band noise, such as RFI. A DMT system has even greater flexibility to deal with multiple RFI tones by simply not assigning information to those sub-channels where RFI exists (see Fig. 7.10). RFI is more likely to be picked up on lines in rural areas where more overhead cabling exists.

7.4.6   CAP/QAM versus DMT

The transmission options described above each have advantages and disadvantages. These have been the subject of considerable debate in the ANSI and ETSI standards forums, both for ADSL and VDSL. It has been recognized that QAM and CAP systems can be implemented in VLSI with similar complexity and achieve similar performance, and therefore the real debate has been between single-carrier (CAP and QAM) versus multi-carrier (DMT) technology. Some of the arguments and counter arguments that have been cited for each technology are listed below (without prejudice):

- DMT's ability to adaptively shape the allocation of information and transmit power across a given bandwidth enables it to better approach optimum performance than is theoretically possible with CAP/QAM;

- the complex start-up procedure required for DMT could result in longer activation times than is possible with CAP/QAM;

- DMT offers greatest performance advantages over CAP/QAM at high transmission rates — the shorter loops at higher rates have more usable bandwidth giving DMT a greater degree of freedom (more usable sub-channels) for flexible adaptation;

- DMT uses block (FFT) processing that results in transmission delay (latency) — this may violate some telco/international specifications for services such as telephony and ISDN at ADSL speeds, whereas it is not an issue at VDSL speeds;

- DMT is more adept than CAP/QAM at coping with multiple RF interferers;

- a high peak-to-average ratio in the DMT transmit signal can lead to clipping noise and expensive analogue-to-digital conversion — part of the power of the error correction may be used up in dealing with this additional noise;

- it is simple for DMT to meet an arbitrary transmit power mask specification to meet spectral compatibility requirements;

- echo cancellation in DMT systems is non-trivial;

- DMT has a greater inherent immunity to impulsive noise than CAP/QAM;

- CAP/QAM can use simpler forward error correction than DMT;

- DMT requires minimal equalization with slower signal-processing rates than CAP/QAM;

- DMT hardware is more easily programmable to support a variety of upstream and downstream data rates with potential for on-line reconfiguration;

- DMT need be no more complex to implement than CAP (in terms of signal processing MIPS per area of VLSI real estate), although it may be more complicated to understand and design, in particular the start-up sequence.

It is left to the proponents of each technology to debate the validity, importance and cost implications of the above statements. In the early days of ADSL development, CAP/QAM prototypes performed worse than DMT but some of the core 'bit-pump' signal processing was more highly integrated. Manufacturers' development programmes for subsequent generations of each technology are providing the incremental steps towards the telco's 'ADSL target' of high performance and low cost. The picture will change rapidly over the next two years or so as more manufacturers enter the ADSL/VDSL market. Increasingly there will be a focus on differentiators such as quality of implementation, added functionality and network management capability.

7.5   ACHIEVABLE TRANSMISSION PERFORMANCE

7.5.1   ADSL Performance

In the presence of a very pessimistic noise source (HDSL and ISDN crosstalk, plus RFI), DMT ADSL is anticipated to be able to operate (with a 6 dB performance 'safety' margin) over customer lines having an insertion loss of up to about 51 dB at 300 kHz. This corresponds to a range of around 3.6 km to 5.75 km depending on cable gauge for 2 Mbit/s downstream transmission with a 16 kbit/s control channel. This level of performance puts in excess of 90% of BT customers within range of 2 Mbit/s ADSL. At this lower end of the range of ADSL data rates, performance of CAP systems has recently become comparable. The performance of this 'basic' system can be increased by the use of additional signal-processing and transmission techniques. For example, the combined use of trellis coding, echo cancellation and optimized transmit-power distribution could increase the performance by around 5 to 6 dB, thus giving even greater reach and network penetration for an equivalent level of performance.

The performance of the base 2-Mbit/s ADSL system reduces by around 2 dB (about 130 to 220 m) if capacity for ISDN is also included in the control channel payload of an FDM implementation. It should be noted, however, that ADSL using DMT modulation and/or interleaved FEC will violate the 1.25 ms ISDN latency requirement. If the 2-Mbit/s payload of the ADSL system is to be transported in ATM cell format, then the ATM overhead (without significant OAM cell overhead) will reduce performance by around 0.8 to 1.8 dB (about 60 to 160 m).

In an equivalent pessimistic noise environment, systems operating with a 4-Mbit/s payload will operate over loops having an insertion loss of up to about 43 dB at 300 kHz [7]. This corresponds to a range of about 3.1 km on 0.4 mm cable or 4 km on 0.5 mm cable. Similarly, systems operating with a 6-Mbit/s payload will operate over loops having an insertion loss of up to about 38 dB at 300 kHz. This corresponds to a range of about 2.7 km on 0.4 mm cable or 3.6 km on 0.5 mm cable. Table 7.1 and Fig. 7.11 summarize the results for basic DMT ADSL in the presence of pessimistic crosstalk noise. As with the 2-Mbit/s system, range, and hence network penetration, can be further increased for the 4- and 6-Mbit/s systems by the use of trellis coding, echo cancellation and optimized transmit-power distribution.

Table 7.1   'Basic' ADSL Performance in terms of insertion loss at 300 kHz.

Insertion loss 0.5 mm Cu	No ISDN No ATM	With ISDN No ATM	No ISDN With ATM
2.048 Mbit/s Payload	51 dB	49 dB	50 dB
4.096 Mbit/s Payload	43 dB	42 dB	42 dB
6.144 Mbit/s Payload	38 dB	36 dB	36 dB

It should be noted that the performance measures in Table 7.1 are in the presence of pessimistic crosstalk and RFI. In the absence of this noise, 2-Mbit/s systems achieve a maximum range corresponding to over 99% of BT loops.

7.5.2   VDSL Performance

In contrast to ADSL, the number of current developments of actual VDSL transceivers is more limited [2]. Hence, most performance estimates are based on simulation with only a few early VDSL prototype measurements having been published [8]. There is a good history of correlation between simulation results and actual implementation of copper-network digital-transmission systems spanning ISDN, HDSL and ADSL technologies.

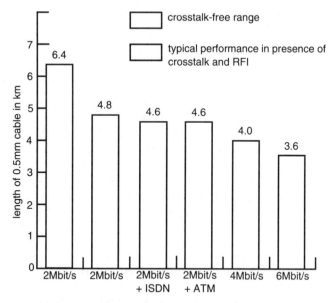

Fig. 7.11   'Basic' ADSL performance in terms of range.

A reasonably pessimistic noise environment for VDSL is to have 12 FEXT interferers. Under these conditions, a 'basic' DMT system operating over 0.5 mm cable with EMC-legal transmit-power level, FDM downstream/upstream channel separation and 8.8 MHz operating bandwidth shows the potential to achieve 25 Mbit/s downstream transmission over 1 km with a 6 dB performance margin. As the VDSL transmitter is moved closer to the customer, such as in a fibre-to-the-kerb architecture, the same basic DMT system could potentially deliver more than 40 Mbit/s over distances up to 300 m.

As with ADSL, the performance of this basic DMT VDSL system can be increased by around 6 dB via the use of trellis coding, echo cancellation, optimized transmit-power distribution and larger DMT sub-channel constellations. Although greater transmit power would further increase performance, the scope for this is limited with VDSL, since transmit power significantly greater than +10 dBm would be prohibited by EMC-emission constraints and worst-case pair balance. This is lower than ADSL transmit-power levels because VDSL operates at frequencies where the copper cable becomes a more efficient radiator. It is technically feasible to use more signal-processing power to extend the range of a 25-Mbit/s or 51-Mbit/s VDSL transceiver beyond the distances quoted above. However, the size and power constraints of externally located VDSL equipment together with the inevitable cost constraints could prohibit the use of more sophisticated systems until future highly integrated generations of transceiver chips become available.

7.6   STATUS OF THE TECHNOLOGY

7.6.1   Trials and equipment

Following measurements on the first ADSL prototypes in early in 1993, ADSL has now left the laboratory and has been used in field trials of video on demand by telcos such as BT and Bell Atlantic. Around 30 other telcos are planning to trial ADSL during the coming year in Europe, the USA, South East Asia and Australasia. These trials are planned to evolve from small-scale technology trials to much larger market trials and are increasingly focused on fast Internet access. ADSL proved itself technically viable in a 60-customer BT video-on-demand trial in 1994. The transmission technology was very reliable and in 1995 BT used ADSL to offer a wide range of interactive services to 2000 customers in a market trial.

7.6.2   International standards

The details of the transmission technology for ADSL have been studied by the American National Standards Institute (ANSI) since 1991. In January 1993, BT, Bellcore and Nynex performed measurements on early DMT, CAP and QAM prototypes. In Florida in March 1993, after considerable debate, ANSI working group T1E1.4 agreed to base its interface definition [4] on the DMT modulation method [9]. The first issue of the ANSI ADSL standard was ratified early in 1995.

The ANSI standard [4] contains an annex which gives details of issues which are specific to 2.048-Mbit/s ADSL requirements, of particular interest to Europe and Australia. The European Standards working group ETSI STC TM3 RG12 was highly instrumental in providing ANSI with information and text for this annex. ETSI has recently started its own work item with the intention of producing a European Technical Report (ETR) on ADSL.

The ADSL Forum is an international group of telcos and ADSL suppliers. It seeks to advance the rapid world-wide deployment of ADSL by facilitating development of interoperable end-to-end ADSL-based network components. Other groups such as the Digital Audio-Visual Interactive Council (DAVIC) are examining the overall network architecture for interactive broadband services. ANSI and ETSI are producing standards and reports covering ADSL chip requirements.

The ADSL Forum will bridge the gap between silicon chips and networks by examining the interface requirements not only for ADSL 'boxes' but also for line-cards.

Both ETSI and ANSI have begun technical studies on VDSL and a VDSL standard has been proposed [10]. The ADSL Forum and DAVIC have also discussed aspects of VDSL.

7.7   THE FUTURE FOR ADSL AND VDSL

ADSL has left the laboratory and has proved its ability to perform in a real network environment as part of video-on-demand trials. The next stages in the evolution of this technology will see manufacturers working to improve performance, functionality and levels of integration as well as reducing costs. VDSL will also follow a similar trajectory from current early prototype developments towards fully engineered and integrated products. International standards bodies will help stimulate and accelerate such developments. For their part, telcos will be working to develop slick operational processes for the planning, deployment and management of these advanced access technologies [11].

Broadband hybrid fibre/copper networks based on optical fibre and VDSL technologies may borrow from the experience gained from the development of telephony over a passive optical network (TPON) [12].

There are numerous access network strategies and applications that are well served by ADSL and VDSL. Various telcos will choose to employ these technologies in different ways depending on their local competitive and regulatory environment.

ADSL provides a per-customer connection in which, for the vast majority of customers, the network infrastructure (copper pairs) is already in place. Capital expenditure to create an ADSL line is matched by a potential new revenue stream. Thus ADSL reduces up-front speculative expenditure by deferring equipment deployment until the customer requests service. This makes ADSL ideal for trials of new services.

Initially, the major trials and applications using ADSL have focused on the residential market with services such as video on demand. However, there is now an increasing trend to consider ADSL for business applications. The ability of standard-compliant ADSL chips to offer bi-directional data rates in excess of 500 kbit/s, together with POTS and the high-speed downstream channel could be attractive for offering data services such as high-speed Internet access, SMDS and frame relay to businesses and teleworkers. The downstream channel, will also be used for remote database access (e.g. for estate agents, sales people, etc). ADSL is useful for offering services having limited appeal within a confined geographical area since it serves customers as opposed to areas [13].

New multimedia interactive services may be rolled out and marketed on a town-by-town basis. Customers requesting these services can be provisioned rapidly with ADSL. Then, when a reasonable cluster of customers in an area have taken up service [13], a shared infrastructure architecture, such as fibre with VDSL copper tails, may be attractive and the telco can be proactive in taking fibre to those customers. The ADSL and VDSL delivery systems could provide the same customer interface.

In relative terms, the capacity of the copper network is limited. However, ADSL and VDSL will facilitate the process of taking new services and optical fibre technology increasingly closer to the home, the first step towards national superhighways.

REFERENCES

1. Waring D L, Lechleider J W and Hsing T R: 'Digital Subscriber Line Technology Facilitates a Graceful Transition from Copper to Fiber', IEEE Communications Magazine (March 1991).

2. Chow P and Cioffi J: 'Broadband Digital Subscriber Lines', IIR Maximizing Copper Conference, London (March 1995).

3. Cole N G et al: 'A low-complexity, high-speed digital subscriber loop transceiver', BT Technology Journal, 10, No 1, pp 72-79 (January 1992).

4. ANSI T1E1.4: 'Asymmetric Digital Subscriber Line (ADSL) Standard T1.413', American National Standards Institute (ANSI), Working Group T1E1.4, (1995).

5. Cook J W and Sheppard P J: 'ADSL and VDSL Splitter Design and Telephony Performance', IEEE JSAC (December 1995).

6. Bingham J A C: 'Multicarrier Modulation for Data Transmission: An idea whose time has come', IEEE Communications Magazine, pp 5-14 (May 1990).

7. Kirkby R H: 'Results of Performance Simulations of Various ADSL Configurations', IEE Colloquium on High-Speed Access Technology and Services, including Video-on-Demand, London, UK (October 1994).

8. Harman D D, Huang G, Im G H, Nguyen M H, Werner J J and Wong M K: 'Local Distribution for Interactive Multimedia TV to the Home', IEEE, pp 175-182 (1994).

9. Cioffi J M et al: 'A DMT Proposal for ADSL Transceiver Interfaces', Amati Communications, Palo Alto, ANSI T1E1.4/92-174 (August 1992).

10. Cioffi J M and Bingham J A C: 'Workplan for Broadband Digital Subscriber Lines Project Evolution', ANSI T1E1.4/95-011 (February 1995).

11. Adams P F, Foster K T and Hind M: 'The operations and maintenance aspects of high-speed copper access transmission systems', ICC94, New Orleans, USA (1994).

12.   Ritchie W K (Ed): 'Local telecommunications networks', Chapman & Hall (1993).

13.   Interview with Ray Smith of Bell Atlantic by D Kline: 'Align and Conquer', Wired, p 110 (February 1995).

8

ON-LINE PUBLIC MULTIMEDIA KIOSKS

J A Totty

8.1 INTRODUCTION

Multimedia kiosks are public-access screen-based systems, providing the general public with temporary access to information, sales or entertainment services, the purpose of the kiosk being to secure from the user a commitment to purchase goods or services, and to provide instant fulfilment where possible.

Kiosks are appearing in retail (including retail finance) outlets and travel sites, but the installed base has remained relatively low, reflecting uncertainty on the part of site owners, often resulting from a lack of established kiosk technology, networking and content management on the supply side. Happily, the position seems to be changing, as there is a growing desire on the part of both content providers and major IT companies to enter kiosk markets as a precursor to the exploitation of larger business and domestic markets.

8.2 SITES AND SERVICES

Kiosk installations can currently be found in a variety of sites, but from a commercial perspective there are only two site types. The first is the private site, where a kiosk sells the site owner's products or services (and little else). Examples of private sites would be banks, building societies, tied car showrooms, shops and supermarkets. Two or more private site owners may of course enter into cross-selling arrangements. The second type is the public site, in which the site owner will allow a broader range of products and services to be sold through the kiosk, examples being airport lounges, bus and coach stations, shopping malls, local authority sites, town-centre pedestrian areas, railway stations and motorway service areas. Kiosks in the latter group are often described as 'multiservice' kiosks.

Typical services offered range from simple (passive) advertising, through catalogue browsing (e.g. browsing information on financial services, holiday destinations or products for sale), to specific searching (e.g. what flights are available on a specific date). The kiosk can provide a telephone, videophone or data link to a service-specific sales desk for the handling of enquiries, and can

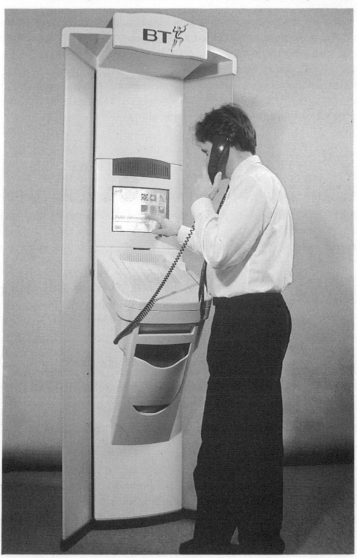

Fig. 8.1   Demonstrator multiservice kiosk at BT Laboratories.

provide the user with hard copy of relevant information (e.g. a map showing the location of a hotel). A 'top of the range' kiosk (see Fig. 8.1) can confirm an order or reservation, take payment for the selected product or service, and provide a receipt for the transaction. Such kiosks are described as 'full function' kiosks. For the purchase of printed information, travel or venue tickets, or for the issue of vouchers, instant fulfilment is possible. Where consumer products or very high value services are being bought, fulfilment requires collection or delivery by external means.

Non-retail services can be envisaged, particularly where there is a need to make central government or local authority services available to the general public. Electronic benefits payment and local authority information are valid future offerings in this context, as is televoting.

A further set of potential services rotates around games and gambling, and the success of the UK National Lottery suggests some potential in this area. Kiosks are already installed in some betting shops.

8.3   KIOSK CONFIGURATIONS

Since the major drivers for kiosk evolution are to be found in retailing, kiosk technology has reflected a desire to emulate traditional selling methods. Specifically, kiosk developers, driven by vendor needs, have attempted to replace, one by one, the four recognizable stages in the selling process, which are:

- to capture the buyer's attention and retain attention during advertising, appealing to both reason and emotion by positioning the product in a lifestyle context, linking the vendor's product and brand with values that are shared by the buyer;

- to provide detailed information, emphasize benefits, offer options;

- to enter into dialogue with the buyer (if necessary), understand the buyer's circumstances, and eliminate objections to sale;

- to close the sale, secure payment or commitment to buy, initiate fulfilment process, and congratulate the buyer.

Kiosk terminals can be grouped into four types, reflecting their function, namely POI (point of information) terminals, browsers, option selectors, and full-function kiosks. POI terminals give information in a predefined order only. Browsers permit free roaming by the user but do not offer any user interactivity beyond navigation, achieving the first two stages of the selling process, but requiring human intervention to continue the sale.

Option selectors allow users to assemble a product or service package according to their liking, and to register an interest by inserting a closed user group card, such as a retail store card, or having a product description voucher printed which can be taken to a sales desk for manual processing, achieving the first two stages and some elements of the third stage in the selling process. Full-function kiosks provide a sales desk channel for the handling of customer enquiries (voice or videotelephony), and a means of handling financial transactions and receipt printing, achieving all stages of the selling process.

In all cases, keyboard interaction is generally eschewed in favour of touchscreen, the combination of active pictorial elements (buttons or icons), and touchscreen being particularly well suited to applications intended for naive users.

It is worth noting that all four types of kiosk have traditionally stored their multimedia content locally, content being prerecorded on hard disk or CD-ROM residing within the kiosk cabinet. As a result, there has been little interest in large-scale networking of kiosks, except for the carriage of transaction traffic (established in the automatic teller machine service) and voice calls stimulated by the fixed content. However, the emergence of suitable networked multimedia authoring and run-time software signals an end to the traditional total dependence on local content. Figure 8.2 charts the evolution of multimedia kiosk technology to date, and shows the emergence of remotely sourced content delivered via network.

In the stand-alone configuration (currently the most common arrangement to be found in live installations) content is held within the kiosk, and no networking is invoked. Stand-alone kiosks are suitable for basic browser applications, particularly where the content has a low churn rate.

The configuration can be extended to provide a low bit-rate data line or voice telephone line to a service-provider back office, allowing sales or service enquiries to be handled, or simple data transfers such as account balance check or goods-in-stock check. The configuration is particularly suitable for option selectors in private sites, and echoes current trends in catalogue shopping and automatic teller machine (cashpoint) system design. For small amounts of content with low churn, it is possible to edit the fixed content from the remote helpdesk site by file transfer.

The on-line multimedia configuration extends the capability of the kiosk still further by holding the multimedia content at a remote server, allowing for high content churn rates, and offering access to a potentially large selection of services, but requiring a trade-off between audiovisual richness of the content and network economics. It is particularly suitable for multiservice kiosks in public sites, or for private sites with significant service variety and churn. Enquiry and transaction lines can be provided, as in the option selector case.

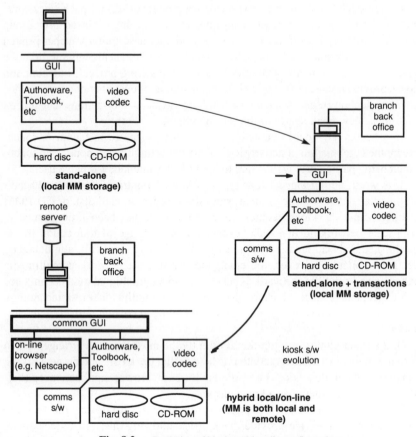

Fig. 8.2   Evolution of kiosk multimedia configurations.

8.4   ON-LINE ARCHITECTURE

It is possible to design scenarios around any one of a number of CPE/network combinations, including TV or PC with copper, fibre, or terrestrial or satellite radio. For BT's kiosk activities in the short term, however, a combination of PC and ISDN networking gives the optimum solution:

- offering excellent geographical coverage (existing network base);

- using ubiquitous terminal hardware (the PC) and peripherals;

- offering synergy with home and office PC information services market;

- using established multimedia formats (e.g. World Wide Web formats);

- using available (and evolving) service-creation software;

- requiring relatively simple server environment;

- supporting telephony, videophony and transaction handling;

- offering service flexibility, anticipating future regulatory frameworks.

The major disadvantage is the lack of on-line richness resulting from a constrained bandwidth of 2 × 64 kbit/s. Nevertheless, suitable on-line multimedia service creation and run-time software packages are available, suppliers being lured by the growth of Internet World Wide Web (WWW) use, and major suppliers are forming alliances to ensure continued development in this field.

Figure 8.3 shows possible relationships between a range of kiosk types and possible multimedia activities in other mass-market areas, and suggests that there may be a spectrum of kiosk applications with high service richness at one end, and high interactivity at the other.

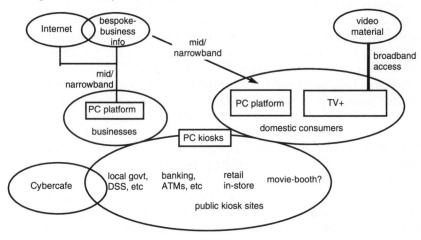

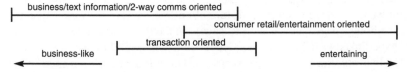

Fig. 8.3   Kiosk applications in relation to business and domestic multimedia.

Figure 8.3 also shows that services requiring high levels of factual content and interactivity for 'getting things done' are clustered around PSTN and ISDN as preferred channels to the business community, typically Internet WWW services and information services brokered by national and international on-line 'supermarkets'. Entertainment services, requiring high levels of media richness, are shown mated with broadband technology as a delivery mechanism.

8.5   MULTIMEDIA CONTENT TRANSPORT

In the stand-alone kiosk, content in the form of text, still pictures (e.g. JPEG encoded) and motion video (e.g. MPEG encoded) is stored on the kiosk PC hard disk or CD-ROM. The content is bound together by hyperlinks which are added during the service-creation process and which control the sequence of activities when a particular screen icon is touched. Multimedia-authoring software packages, such as Authorware or Toolbook, are supplied with built-in run-time functionality, the same software package being used both to create the navigation code (the hyperlink 'glue') and to play back the finished product. To drive a printer or card reader, the navigation code points to a process file which contains the necessary instructions to drive a particular peripheral.

In the case of on-line multimedia, two methods are employed for transferring content from the remote server to the kiosk, namely bulk file transfer (block downloading of entire service scenarios) and on-line browsing (piecemeal downloading of one or two pages at a time to PC RAM-cache).

Although bulk file transfer is practised by at least one kiosk supplier, the technique requires broadband networking to update content within reasonable times, although it does allow video-rich scenarios to be downloaded, and makes kiosk configuration management relatively straightforward. Block downloading via ISDN2 is time consuming (a three-minute, full-screen, MPEG video scenario could take up to 40 minutes to download) and presents scaling problems with growing kiosk populations.

On-line browsing is favoured by a number of kiosk manufacturers since this is the mode of operation of the growing Internet WWW service, and browser software such as Netscape or Mosaic is readily available, allowing access to hyperlinked content usually in the form of text and still pictures, formatted in hypertext mark-up language (HTML) and loaded on to a relatively simple Unix or similar server. One interesting feature of on-line browser software is the ability to compile a page (such as a menu page) 'on the fly' by assembling the necessary hyperlinked visual images (e.g. icons) on to a page grid in real time.

On-line browser software packages are not able to handle real-time video content at this time, though Java™ and Shockwave, for example, offer some improvement in richness. It should be noted that it is not possible to deliver full-screen MPEG video in real time over ISDN2 at all, although H.261 video (lower motion fidelity) can be delivered using the cards and software from PC videoconference systems. However, the position for ISDN2 networking is that full-motion video sequences, where required, have to exist as locally held content, which implies that site visits will be an element of service creation and churn management, unless the sequences can be so short as to permit scalable downloading to kiosks.

A number of kiosk manufacturers have developed hybrid kiosk software, which fuses together the fixed content tools (e.g. Authorware) and on-line browser tools (e.g. Netscape) under a seamless graphical user interface. For private sites with only one or two main services, this is an ideal solution. For multiservice kiosks where churn is high, the options are clear — content must either be created for browsing only, or else the logistics associated with holding video content locally have to be planned in advance and costed into the service creation process.

8.6 RICHNESS VERSUS INTERACTIVITY

This picture of on-line and supplementary fixed content is explored in Fig 8.4, which positions a number of multimedia services in terms of the content richness and level of interactivity offered. Richness in on-line services is a function of available bandwidth, but can be boosted by the local addition of supplementary (video) content. Interactivity is a function of the kiosk-content navigation design, the peripheral-equipment level and the provision of communications links and transaction handling systems.

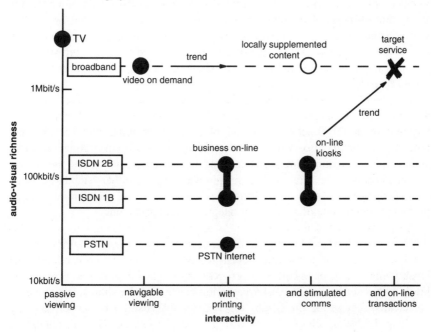

Fig. 8.4   Richness versus interactivity for a selection of services.

Since the *raison d'etre* of kiosk services is fast turn-round, the combination of high interactivity and intermediate richness is an appropriate solution. In fact, for most kiosk applications, video sequences lasting more than two minutes will have a counter-productive effect. Crisp, imaginatively designed pages and good audio can be just as effective as video, despite the power of full-screen video to reach users at an emotional level or to explain abstract concepts. The key feature of any useful kiosk service, then, is interactivity.

8.7   KIOSK INTERACTIVITY

Kiosk interactivity includes the following three components:

- the degree of user contol over the content displayed (navigation);

- the terminal peripheral equipment level (printers, etc);

- provision of sales-desk links and financial transaction handling.

The degree of user control over the content displayed varies considerably across multimedia applications. For video-rich applications, a tape-transport analogue is often used, offering 'play', 'stop', 'review' functions. For the reasons already outlined, this mode of control has little value in kiosk environments, and the preferred mode of operation is via hyperlinked icons (pictorial or labelled button menu items) to successive pages or screens, allowing the user to quickly funnel into the product or service of interest. The need for quick turn-round of business suggests that menu trees on kiosks should be shallow (no more than five layers deep) and self-explanatory.

Terminal peripheral equipment levels vary considerably across terminals according to the target application, but to achieve the level of interactivity required on a full-function terminal, plain-paper printing, voucher printing, receipt printing (local or with on-line validation), cash (coin) validation, magnetic stripe card reading and smart-card reading are all necessary in some combination. Fortunately, most of these peripherals are either standard office equipment products or used in point-of-sale products, and the terminal design problem becomes one of component integration and packaging.

The real power of the kiosk is in its ability to stimulate calls to sales help-desks or advertisers, and to handle financial transactions. Both require networking, but local-line contention may occur if a browsing instruction is made at the kiosk while a sales-desk call is in progress. A simple solution is to provide the kiosk with a separate sales-desk/transaction channel in the short term, allowing voice calls to be set up directly by touching an advertising icon on screen or by touching the 'sales helpdesk' icon within a specific service. The sales helpdesk channel can be extended to link the kiosk to the sales back-office data system to permit price checking or transaction handling.

8.8 PAYMENT METHODS

Any kiosk with a line to a remote sales desk can provide financial transaction handling by the (insecure) method of having the user quote their credit card number over the voice channel. For the purchase of goods, fulfilment requires that goods are delivered to the credit card owner's registered address, which goes some way to overcoming fraud problems. The use of magnetic stripe cards in (most) unattended terminals is currently restricted by card issuers. However, the introduction of smart-card payment technology should permit direct transfers between unattended kiosks and sales-desk back-office systems in the near future, smart cards being relatively resistant to cloning fraud. During the transition phase between magnetic stripe and smart-card use, hybrid card readers will be required. Such readers are already becoming available and will be procured in volume for payphone products.

At this time, two types of smart-card are of interest, namely the token card and the electronic purse card. The token card is a prepayment card which allows units of credit to be spent on predetermined services only. Examples are the pre-paid phonecard and parking cards. When the credit is spent, the card is either thrown away or topped up by a recognized dealer. The electronic purse card, e.g. Mondex, is in essence electronic money, which can be spent or accumulated in any way (including from person to person) provided the funds' transfer electronics exist. During a transaction, the digital equivalent of pounds and pence is carried over the network, and considerable effort has been expended on encryption to make fraud (almost) impossible. The electronic purse card offers the prospect of secure teleshopping without the need for service subscription and is an ideal future payment mechanism for kiosk services.

8.9 KIOSK MANAGEMENT

For private kiosk systems running on LANs and accessing a local server, system management is a modest extension of standard office-automation practice, the added complexity being the need for content creation and installation. For large, geographically distributed systems working over the public network, service management requires significant effort. While undoubtedly adding cost to the system, service management should also be seen to have revenue potential.

Figure 8.5 shows a potential national architecture for a large kiosk system, and illustrates the need for management processes and systems in the server, network, and kiosk terminal domains.

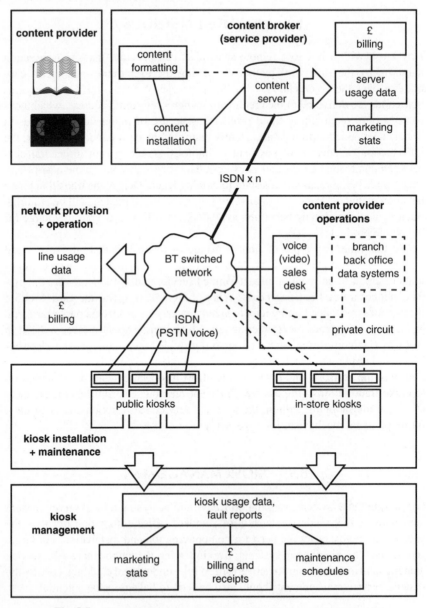

Fig. 8.5   Networked kiosk architecture and core service-management system.

Surrounding the core service-management systems, and not shown in Fig. 8.5, are the traditional service operational-support activities including provision, maintenance and repair, helpdesk and content management.

Kiosk and content usage data can be translated into marketing statistics which allow content providers, service providers (brokers) and kiosk operators to assess how well specific services perform in certain site types or geographical areas. The usage data also provides the basis for billing and reconciliation activities. From the terminal operator's point of view, kiosk alarms and fault reports need to be processed quickly, linking in to existing automated maintenance systems where possible.

8.10   FUTURES

The future of kiosks is probably best suggested by assuming that technologies will evolve to meet pressures for more richness and interactivity across growing mass IT/communications markets.

A small increase in richness may arise as browser software evolves in the short term. In the longer term, richness may be met over restricted bandwidth channels by the transmission of a 3-D image to the kiosk and its subsequent animation on screen, but a more likely scenario is that kiosk architectures will yield to broadband networking growth, allowing more service 'presence' at the kiosk.

Opportunities may arise in evolving electronic payment fields, particularly smart-cards, eliminating the need for user subscription, and allowing spontaneous interaction. Cards holding user identification data will allow individual spending patterns to be tracked, leading to precise targeting of new product or service offerings. A coming together of multimedia and smart-card technologies may bring simpler architectures, allowing high-value transactions from desktop PC or kiosk.

The scope of kiosk services will grow to encompass service customization, electronic benefits transfer, televoting, gambling, and arcade-style games. In the longer term, content granularity will change as individuals assert their right to become content providers.

9

SUPPORTING TELEWORK-ING WITH MULTIMEDIA

M J Gray

9.1   INTRODUCTION

Teleworking is a flexible way of working which covers a wide range of work activities, all of which entail working remotely from an employer, or normally expected place of work, for a significant proportion of work time. Teleworking may be on either a full-time or a part-time basis. The work often involves the electronic processing of information, but always involves using telecom-munications to keep in contact with the remote employer.

This definition excludes the traditional 'outworkers', as well as people who work at home for a day now and then, but includes a wide variety of working situations:

- people working at home (e.g. programmers);

- people working from home (e.g. salespeople);

- people working at work centres (such as telecottages, or satellite offices).

Major companies are showing increasing interest in the use of teleworking because of the benefits it can bring to the company and its employees. The company benefits include some (but not usually all) of the following:

- better customer service through increased flexibility in the workforce;

- increased productivity, often due to less sick leave, less interruptions and lack of background office noise;

- skills retention and improved recruitment opportunities;

- reduced overheads, usually through savings in office accommodation;

- resilience to power cuts, bomb threats and public transport problems.

However, in order to get the best from a teleworking programme, careful consideration needs to be given at the planning stage to a wide range of issues.

- Management issues

  — selection criteria;

  — training;

  — security;

  — supervision;

  — trade unions;

  — terms and conditions;

  — support at home.

- Legal considerations

  — health and safety;

  — planning regulations;

  — capital gains tax;

  — income tax;

  — insurance;

  — mortgage and tenancy agreements;

  — local council taxes.

- Human aspects

  — home environment;

  — social isolation;

  — child care;

  — career advancement.

Technology will have little or no impact on many of these issues, but in other, important cases it can be used to enhance benefits or overcome difficulties. A key area is the last of the 'management issues' listed above — 'support at home'. It is not enough merely to provide a homeworker with the capability to do their normal work at home; in recognition of the remoteness of the worker, adequate support must be given. Support is the key to successful teleworking. In order to establish how much support is 'adequate', BT Laboratories has been involved in a number of teleworking projects. Of these, the one most directed at investigating the support issue, and the most well-known, is 'The Inverness Experiment' [1].

9.2   BT'S INVERNESS EXPERIMENT

The purpose of carrying out teleworking experiments is to explore, in detail, particular working situations and to try out ideas for providing teleworking support using 'leading-edge' technology.

As with any experiment, it was important not to change too many variables at once. For this reason, and to allow direct comparison, the work patterns of the operators working at home mirrored those of the Directory Assistance Centre-based operators. In a pilot or trial this constraint would not exist and increased efficiency and flexibility could be achieved through adjusting the working hours of the teleworkers.

The objectives of this teleworking experiment were:

- to demonstrate that call-centre operators can successfully work at home;

- to explore how the facilities and support provided for office-based workers can be extended to homeworkers;

- to investigate how the technical and non-technical problems of teleworking can be overcome;

- to look at how support for 'human factors' and supervisory issues can be provided for teleworkers using technology.

9.2.1   The teleworking system

The experimental teleworking system for the Directory Assistance (DA) operators was developed at BT Laboratories. It was designed so that the job of the teleworking operators remained essentially the same as that of the Centre-based operators. The experiment involved ten BT operators in the Inverness area, where it was possible to take advantage of an extensive ISDN network.

The system was also designed with several types of user in mind and these were the main participants in the teleworking experiment:

- the teleworkers themselves;

- supervisors within the DA Centre;

- senior operators in the DA Centre with responsibility for various miscellaneous clerical and administrative functions;

- centre-based operators.

Each of these groups interacted with the system via a terminal with each terminal being tailored, in terms of functionality, to meet the requirements of the group it served.

The teleworking system incorporated several support facilities for the teleworkers, based on an extensive study of the needs of the operators. These included:

- videophone — for dealings with supervisors, and for socializing with colleagues when off-duty;

- e-mail — allowing messages to be sent between teleworkers and between the DA Centre and teleworkers;

- electronic noticeboards — for shared information;

- electronic forms — to replace the paper forms used by operators in the DA Centre;

- newsflash — to 'broadcast' urgent information to the homeworkers;

- SOS — for notifying the supervisor of a domestic emergency;

- comfort break — for requesting a comfort break during a shift.

Figure 9.1 shows the normal configuration of equipment in the Inverness Directory Assistance Centre as used by the operators based in the DA Centre (see Fig. 9.2).

Incoming calls from the customers were distributed to the operators by the automatic call distribution (ACD) switch which gave an even loading among the operators and equalized the call-waiting time. The operator answered the call from a console, which consisted of a desktop unit (also known as a 'turret') and headset, and then used a special terminal to interrogate the directory database and find the telephone number required by the customer. All of the components shown in Fig. 9.1, apart from the directory database, were located in the DA Centre.

These essential elements, which allowed the operators to do their work, were extended into their homes using ISDN as shown in Fig. 9.3.

However, from the initial feasibility study it was concluded that additional support was essential for remote workers. When the additional support facilities

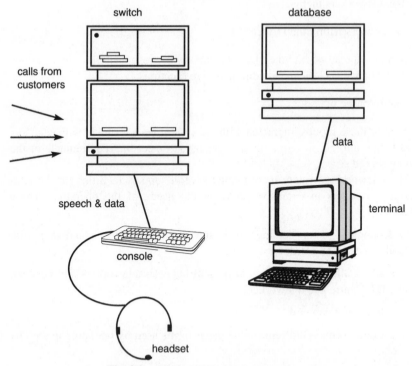

Fig. 9.1   Normal DA Centre-based configuration.

for the experiment were added, the situation became more complicated. Figure 9.4 illustrates the configuration of equipment which was used to support the tele-working operators.

The support facilities were provided largely by the central support-services computer. Multiplexers were used to combine the data and voice channels for transmission across one of the two 64-kbit/s ISDN channels. The videophone, which appeared in a window on the teleworker's screen, used the second channel. The teleworker's terminal was a personal computer which emulated the 'Centre' terminal (as shown in Fig. 9.1). The computers all used a Unix operating system. All the user interfaces used X Windows Motif.

There was a supervisor's terminal, located in the Centre, from where the supervisor could manage and interact with the teleworkers. A second supervisory terminal was provided for the use of the Senior Operator responsible for certain clerical tasks within the DA Centre.

There was also a terminal, with limited functionality, provided in the Centre's welfare room — where the Centre operators usually sit during breaks from work — to allow socializing between the teleworkers and Centre operators.

Fig. 9.2   An operator in the DA Centre.

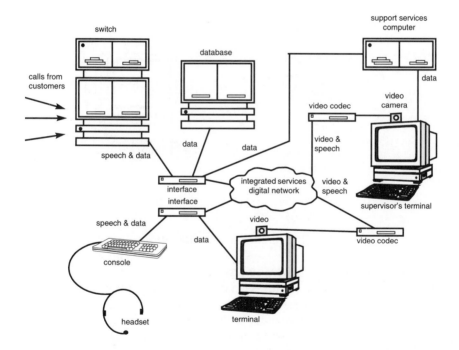

Fig. 9.3   Remote working.

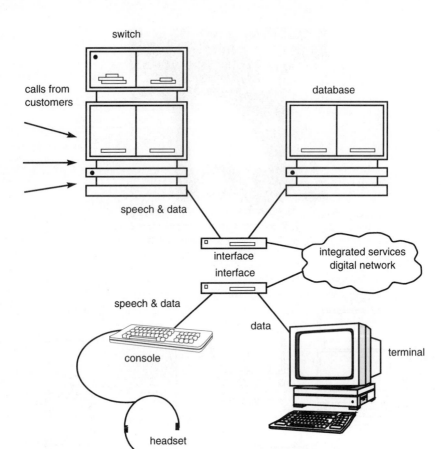

Fig. 9.4   Remote working with support.

9.2.2   General considerations

Teleworkers and supervisors were each given 2-3 days training on the use of the teleworking system, prior to the beginning of the experiment in June 1992. Throughout the 12-month experiment the teleworking system was monitored with data being gathered on the usage and perceptions of the support facilities. In addition, the Department of Psychology at the University of Aberdeen tracked the psychological effects on behalf of BT Laboratories.

9.3   EXPERIMENT RESULTS

9.3.1   Teleworkers

The operators who took part in the experiment benefited from savings in time and money through avoiding commuting, reduced stress and greater flexibility. Drawbacks were relatively minor, with the biggest problem being the waiting time for faults to be fixed (see also section 9.3.4).

Even with these minor drawbacks Fig. 9.5 shows that the teleworkers were generally well satisfied with teleworking throughout the experiment.

In Fig. 9.5 the Y-axis represents a seven-point scale where 7 represents 'very satisfied' and 1 ' very dissatisfied'. The majority of scores are between 5 'slightly satisfied' and 6 'satisfied'.

Figure 9.5 shows one exceptional score (4 on the chart represents 'neither satisfied nor dissatisfied') around week twenty, which was the Christmas period. This was mainly due to resourcing difficulties in the DA Centre over the Christmas period, leading to teleworkers being told not to use the videophone to make contact with the DA Centre before 'logging-on' at the start of a shift. Instead they were told to contact a Senior Operator using the telephone and, as a result, did not receive their daily team talks from the supervisor; the teleworkers missed their daily 'face-to-face meeting' with their supervisor (see Fig. 9.6).

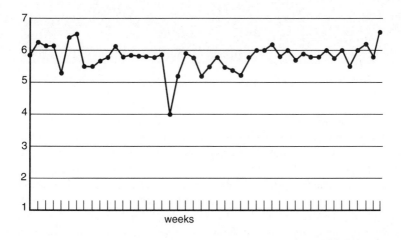

Fig. 9.5   Response to the question: 'Overall how satisfied have you been with teleworking this week?'

Fig. 9.6   One of the teleworkers at work.

For the majority of teleworkers, the expected problem of isolation did not materialize. In part, this is due to the range of communications facilities that they had at their disposal, but their experience also seems to back up findings, reported from other teleworking programmes, that isolation is largely a perceived problem. For teleworkers who are part of a well-planned teleworking programme, isolation is not a big problem. At the end of the twelve months, the majority of teleworkers had found teleworking so beneficial to them that they wanted to continue teleworking. Nine out of ten teleworkers also indicated that their families preferred them teleworking.

9.3.2   Company benefits

The experiment has shown how teleworking can positively affect the service offered to customers and the productivity of operators:

- flexibility — teleworkers were more willing to work outside their standard hours;

- resilience — teleworkers were unaffected by weather and travel problems;

- skills retention — one operator moved 150 miles away, but continued as part of the experiment;

- productivity — statistics showed that teleworkers were just as productive when working, and they also tended to take less time off through minor illness.

9.3.3   Technology

Videotelephony was an extremely valuable part of the teleworking system. The videophone was used the most out of all the support facilities and was considered, by the teleworkers, to be the most important, easy to use, enjoyable and easy to learn. It was also the major form of communication that the supervisors used to contact the teleworkers. Some of the operators that worked at home suggested that the videophone would have been used even more if it had been accessible while the rest of the teleworking system was switched off.

A videophone was provided in the welfare room at the DA centre but it was used very little — this was due to a variety of reasons, including the difficulty of the teleworkers knowing who would be in the welfare room at any given time. However, the main reason seems to have been the reluctance of the Centre-based operators to use the videophone terminal. This was mainly caused by the operators feeling inhibited, possibly through lack of confidence, about using the terminal in front of their colleagues.

E-mail was also seen as important and useful even though the 'off-the-shelf' package used in the experiment had a poorer user interface than the other facilities, and was therefore not easy to use or learn. After a fairly slow take-up by the teleworkers, and once the system had been modified to make it easier to access and use, e-mail became an effective and essential means of communication for the teleworkers.

The videophone and e-mail were two of the two-way communications facilities used in the experiment, the others being telephone and post. Figure 9.7 shows the average number of times that these facilities were initiated by the individual teleworkers and from the terminals used by the supervisors and senior operators.

Of the other support facilities the electronic form and electronic noticeboard facilities were thought to be the next most important after videophone and e-mail.

Overall the system was ranked highly by the teleworkers. When interviewed at the end of the experiment, the teleworkers indicated that there was little of any significance that they would change on, or add to, the system in order to improve it.

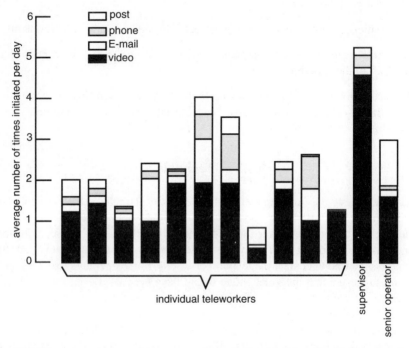

Fig. 9.7   Average usage of the two-way communications facilities offered by the telework system.

9.3.4   Lessons learnt

9.3.4.1   Maintenance

The importance of good maintenance procedures was highlighted during the experiment. In the event of a fault, teleworking operators do not have the option, that centre-based operators may have, of moving to a spare terminal. Instead, the fault must be repaired as quickly as possible, or a decision taken for the operator to travel to the DA Centre. To enable swift repair of the fault, the procedures should be such that the fault is quickly diagnosed and focused action taken to repair it.

For the experiment, a good maintenance programme was arranged, with a guaranteed call-out time for service engineers of not more than four hours (comparable with BT's 'Total Care' maintenance package). However, for part-timers, four hours is a complete shift. Any employer introducing teleworking for people whose work is dependent on the availability of 'information technology' must consider the balance between the cost of fast maintenance response and the wasted time of an employee unable to work.

9.3.4.2   Supervision

The supervisor/teleworker communication channel is an important one, important both in terms of supervision and because the supervisor is the main point of contact for the teleworkers. During the experiment supervisors needed to spend more time giving daily team talks to the teleworkers. Team talks were given using the videophone and therefore had to be repeated for each teleworker. Multi-point videotelephony would overcome this problem.

9.3.4.3   Socializing

For the majority of the teleworkers, the expected problem of isolation did not materialize. In part this is due to the range of communications facilities that they had at their disposal, but their experience also seems to back up findings, reported from other teleworking programmes, that isolation is largely a perceived problem. Potential teleworkers expect isolation to be a problem, but find in practice that it is not. This is exemplified in the experiment by the fact that the teleworkers preferred to use free time, such as regular breaks from work, to 'do their own thing' rather than socialize over the network.

9.4   EXPANDING THE TELEWORKING ARENA

The Inverness experiment showed that teleworking was possible, and beneficial, in a call-centre environment. BT has followed up on the success of this experiment by introducing teleworking in its own call centres. In 1995, one sales area started the ball rolling with 12 people on the '152' service based around the Southampton area. The principles of support, and of the technology used, followed closely the Inverness pattern, the major difference being that all the equipment and software was available 'off-the-shelf'.

The experiment stimulated interest in the possibilities among a number of companies who have 'telephone enquiry agents' working in call centres. BT is now able to provide consultancy on all teleworking issues to help such companies introduce a teleworking programme.

However, the principles of support developed for Inverness can be applied to a much wider range of jobs, although the technology used to give that support may vary. For example, for professional teleworkers, such as the people at BT Laboratories, or colleagues in marketing and sales, the screen-sharing facilities of BT's VC8000 Multimedia Communications Card could be invaluable. Other 'groupware' products, like Microsoft 'Windows for Workgroups' and Lotus 'Notes', will help distributed teams work as one unit.

The time has arrived for teleworking. Companies need its benefits for competitive advantage; individuals desire the freedom and flexibility it can give them. The technology to support teleworking is well established, and getting more extensive by the day. Only old-fashioned management attitudes hinder the growth of teleworking.

REFERENCE

1.   BT Report: 'The Inverness Experiment', BT Laboratories, Martlesham Heath, Ipswich, IP5 7RE (1994).

10

DESKTOP MULTIMEDIA

T Midwinter

10.1   INTRODUCTION

Multimedia has become a commonly used term within the computer industry. Every week there seems to be a new multimedia product launched by one computer company or another. The term 'multimedia' in the context of these products has become synonymous with CD encyclopaedias and stunning graphics performance, but has had little real impact on most businesses. 'Multimedia communications', on the other hand, is a very new topic that has just started to be addressed by the computer and communications industries. Multimedia communications promises to change radically the way business operates and in the longer term is likely to have as major an effect on the fabric of our society as the Industrial Revolution and the invention of the telephone.

Desktop multimedia is rapidly becoming the latest term to escape from the computer industry's marketing department. It is a term used to describe the amalgamation of 'multimedia communications' with the PC on a user's desk. The resultant desktop multimedia terminal consists of a standard PC (configured for the user's normal business) with additional functionality to allow it to communicate with a telephone network. It will allow the user to make calls to anybody around the world, using existing telephone networks. If the other party in the call is capable of video and/or data communications, then the terminal will automatically set up a call to the full capabilities of the two ends, without any further user intervention. One terminal on the desk will provide all the communications needs of the user, allowing the existing telephone, facsimile machine and data modem to be thrown away. One line, one terminal and one unified service — that is the 'multimedia communications' vision, and, as will be shown in this chapter, it is well on the way to becoming reality.

Desktop multimedia is much more than just a technology to allow users to connect their PCs to the telephone network and transmit speech and video at the same time as data. It is an enabling technology that will allow businesses much more flexibility in balancing the needs of their customers and workforce. The customer needs are better met as the business can call on specialists when required. The workforce needs are better met as people will have much more choice regarding where they live, and what hours they work. In essence, the technology promises to break down many of the barriers that distances (and hence travel time) currently impose. If you can ring any colleagues at will, speak to them, see them and discuss, modify or transfer any document or other data with them, then travelling into the office becomes much less necessary. If all these functions can be performed within a multiparty conference, then many people will be able to work from home rather than having to commute. Now include the problems faced by multinational companies, as they try to get international teams of experts together to discuss an issue, and the advantages of working together efficiently, despite the physical distances, become obvious. Desktop multimedia provides one of the base technologies required for all of these work styles — consider the following job example.

- Telephone operator [1] — modern digital telephone systems will allow calls to be answered as easily using a terminal at home as using a terminal in a central office. The telephone system can present a call to the operator's home number and hence the job could be done as efficiently from home as from a central office. The advantages are obvious — zero travelling time, easier emergency cover provision, less centralized accommodation, etc. The major disadvantage would seem to be that the operator could become rather isolated from colleagues, and lose the feeling of company loyalty. However, trials have proved that the use of video as well as audio in conversations allows teleworkers to interact as part of a tightly knit community (see Chapter 9). Although this inevitably incurs extra call charges, the benefits easily outweigh the costs.

Although the example chosen is well understood and documented, it can easily be extended to include most other office-based jobs.

Desktop multimedia has recently become possible due to the merging of many development strands. As well as the technology, standards have played a key role. Over the last ten years, standards for videoconferencing and multimedia have been developed to achieve the vision of global interconnect, at the basic capability level of both terminals. These standards, together with the remorseless march of technology, are now beginning to deliver the dream — one terminal connected via one telephone line across the world-wide telephone system to deliver a simple-to-use, consistent service from telephony at one end of the scale to desktop multimedia at the other (see Fig. 10.1).

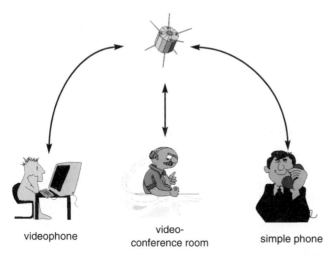

Fig. 10.1   World-wide interconnect.

10.2   A GENERIC DESKTOP MULTIMEDIA SYSTEM

A desktop multimedia system can be considered as two items, all the add-ons to a PC and the PC itself. This chapter will look at a general model of a desktop multimedia system that is capable of connection to an ISDN basic rate interface (ISDN BRI [2]) network (see Fig. 10.2).

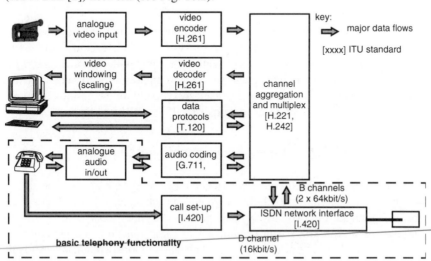

Fig. 10.2   Generic design of a desktop multimedia system.

If a desktop multimedia system is to replace the user's existing telephone, it must be capable of operating like a standard telephone, even with the PC switched off, as well as a full multimedia videophone when the PC is running. Similarly, once powered, the PC should be capable of communicating with other data devices, like facsimile and modem systems. The functionality for these is given below.

10.2.1   Basic telephony functionality

To perform as a simple telephone, when the PC is not powered, puts considerable limitations on the audio unit (telephone) of a desktop multimedia system. However, if the user's normal phone is to be substituted by the desktop product then these limitations must be overcome. The major features are:

- the audio unit must operate even when there is no power to the PC, i.e. all the basic telephone functionality must be line powered or powered from a plug-top PSU;

- the audio unit must be capable of originating and answering calls;

- the audio unit must be capable of operating in an office environment, i.e. a handset is required — in some cases loudspeaking and headset operation would also be useful;

- the audio unit must also be capable of being controlled directly by the PC, when it is powered up, to allow calls to be set up and answered by the PC directly.

Hence the audio unit looks similar to a telephone and in a simple videophone application could be a standard telephone. However, in a desktop multimedia application the telephone needs to be tightly integrated with the PC, to allow the PC and phone to interact during call set-up, call answer, etc; consequently, the audio unit is usually a custom unit designed specifically for the desktop multimedia system. The audio unit needs to include the ISDN BRI interface.

10.2.2   Videophone-specific functionality

To perform as a standards-compliant videophone, the system needs to be able to compress and decompress both the video and speech signals. It also needs to be able to aggregate the two 'B' channels presented by the ISDN BRI together to make a single 128-kbit/s data stream. All of these functions are governed by standards, which will be discussed briefly here and in more depth in the standards section of this chapter. The major features of the videophone-specific functionality are:

- video encode and decode to ITU-T Recommendation H.261 [3];

- audio encode and decode to ITU-T Recommendation G.711 [4] (Standard Telephone Quality 'A'-Law/'μ'-Law Coding);

- optional audio coding to ITU-T Recommendation G.728 [5] (Low Bit Rate, Telephone Quality Coding) and/or ITU-T Recommendation G.722 [6] (Medium Wave Radio Quality Coding);

- channel aggregation and multiplex to ITU-T Recommendation H.221 [7] and ITU-T Recommendation H.242 [8] (aggregation of the two 64-kbit/s channels together to make a single 128-kbit/s channel and multiplex audio, video and data together into that single channel).

The videophone-specific hardware in a desktop multimedia system can be powered from the PC as it does not need to operate when the PC is not operational.

10.2.3   Desktop multimedia-specific functionality

The differences between a desktop multimedia system and a standard videophone are based around the PC integration. The major features of the desktop multimedia-specific functionality are:

- PC application software to control all the telephony functions, video windowing, facsimile, modem and any other functionality provided;

- PC application software to transfer data from one party to another using the ITU-T Recommendation T.120 [9] protocols — these applications include file transfer, shared whiteboard, application sharing, multipoint conference control, network database control, etc (this data will need to be inserted into the data channel for transfer to the remote party(ies));

- facsimile and modem algorithms;

- video windowing hardware may need to be added if the host PC does not include it.

10.2.4   The complete desktop multimedia system

When all the modules described above, and a camera, are integrated together, a complete desktop multimedia system is produced.

The system discussed so far has not made any distinction between add-on functionality, and software running on the host itself. This is deliberate, as the split between host and accelerator card is continually changing. The functions listed above are generic and apply to all systems. Their implementation is product specific, creating the cost/performance differentiation between manufacturers.

To highlight the differences, two products are considered — a hardware accelerator product designed to run on most Windows™-capable PCs, and a software-only product which requires a top-end PC.

- Case study 1— the BT VC8000 multimedia videophone (using hardware acceleration)

  The VC8000 is in many ways the traditional videophone design, in that it does not require much support from the host PC. All of the functions described for the generic videophone are performed on the accelerator card, leaving the PC free to run Windows, user applications and multimedia applications. The processing power available on the accelerator card is approximately three times that of a top-end Pentium processor, which allows the card to produce good-quality video pictures, implement all of the audio-coding algorithms (including the computationally intensive G.728 [5]) and provide T.120 [9] and ISDN interface support.

  The hardware provided in a VC8000-type kit includes a telephone (which operates with the PC power off), a camera, an ISDN interface, a windowing interface and all the processing to implement the video, audio and data coding. Irrespective of whether a call is in progress, the processing power of the host PC is not compromised by the operation of the videophone.

  Hardware acceleration tends to offer the customer a high-quality product that makes far fewer demands on the host PC than the 'software only' solutions. The cost of this type of solution is also greater than a 'software only' solution, but the differential can be much smaller than expected as even the 'software only' system requires some added hardware.

- Case study 2 — a software-based system (no hardware acceleration)

  A 'software only' codec may initially appear to be a very cheap solution, but is, in practice, still quite expensive as it requires hardware which includes a camera, an audio sub-system and an ISDN interface. Some of these features are currently provided on top-end multimedia PCs, but it is not clear that they are really suitable for a full telephony-quality product. In a large office, would a loudspeaking PC really be acceptable in place of a standard telephone?

  The software codecs developed so far have relied on an ISDN card, which also provides PCM audio to the PC and a digitizing card, with camera, to provide a video source. The H.261, H.242 and H.221 modules are performed by the PC in real time. The PCs used for this type of codec are top-end 486s

(66 MHz) or Pentiums, and the videophone operation takes up most of the processing power of the PC. Applications run slowly, or the video stops when both video and applications are active at the same time.

Another major problem with all software-based codecs is that they require the PC to be switched on even to operate as a basic telephone. Consequently, if the PC is switched off, or the application shut down, the ISDN telephone will effectively be unplugged and calls cannot be made or received — not a real videophone solution. In the future there will be phone add-on kits for PCs which will allow basic telephony even with the PC unpowered.

10.3   INTERNATIONAL STANDARDS AND INTERWORKING

The ITU-T, previously known as the CCITT, has been responsible for setting telecommunications standards for many years. During this time it has produced recommendations for all forms of communications equipment ranging from basic telephones, facsimiles and modems up to trunk-signalling equipment and telephone exchanges. In conjunction with PTTs and manufacturers from around the world the ITU-T has also defined sets of standards for desktop multimedia systems. The umbrella standard for multimedia terminals on the ISDN is H.320 [10]. For multimedia terminals on the PSTN, the recommendationis H.324 [11]. One of the features of ITU-T standards is that they take a considerable amount of effort to get correct, but once ratified they can be upgraded, while still maintaining backwards compatibility. This leads to families of products from different manufacturers which interwork with all versions of the standards and removes much of the very costly 'designed for obsolescence' associated with most of today's 'Industry Standards'. The major components in H.320 are given below.

10.3.1   H.261 — the video coding standard

H.261 is the ITU-T standard for real-time video coding and defines three main areas within the video-coding process.

- Video format

    The video format chosen proved to be a very contentious issue when H.261 was developed, as the European requirement was for a PAL-based format and the American/Japanese preference was an NTSC-based format. The final format was a compromise which was acceptable to all countries and is the Common Intermediate Format (CIF). There has also been defined a lower resolution video format called Quarter CIF (QCIF). As all H.261-

based systems use the same video formats, they all interwork together irrespective of whether the originating country operates to the PAL or NTSC TV standard.

- Video decoder

  The standard defines how a reference decoder will work, but does not completely define how an encoder works. At first sight this might seem a bit strange, but is actually very important, as it allows manufacturers considerable flexibility in the encoder design and control. This allows each manufacturer to develop a unique encoder design, thus providing much needed product differentiation between systems.

- Error correction

  A BCH (511,493) error corrector is defined in the standard. This corrector takes 493 bits of data and adds 18 correction bits, to make up a 511-bit frame, for transmission to the far end. The decoding terminal strips out the correction bits and then has three choices, do nothing (throw the bits away), implement a burst corrector which corrects any single burst of up to 6 bits in the 511-bit frame, or implement a random error corrector which can correct any two errors in the frame. Most real systems opt for one of the two correction schemes. Which scheme is best depends on the actual error characteristics of the line, and, in practice, there seems to be little to choose between the two schemes.

10.3.2   G.711, G.722 and G.728 — the audio coding standards

Three different audio standards are defined within H.320.

- Telephony quality speech, 'A'-Law and 'µ'-Law

  This standard is mandatory and all terminals which claim H.320 compliance must include at least G.711 support. However, this is not very onerous, as G.711 is the same standard as is used to define normal telephony-quality speech. In Europe, the 'A'-Law version of the standard is used and in the US and Japan the 'µ'-Law version is used. G.711 supplies 3.4-kHz analogue bandwidth using 64 kbit/s of channel capacity.

- Low-bit-rate speech

  This standard is optional, although most serious terminals include it. The major exceptions are the software (PC) codecs which do not have the processing power to perform G.728. This standard gives 3.4-kHz analogue bandwidth (telephony quality) using 16 kbit/s of channel capacity. Using G.728 it is quite practical to conduct a multimedia call including speech, video and data over a single 64-kbit/s ISDN channel.

- Wideband audio

  This standard is also optional, and is the one favoured by high-quality speech terminals. G.722 supplies 7-kHz analogue bandwidth (approximately medium-wave radio quality) using 56 kbit/s or 48 kbit/s of channel bandwidth.

Videophone terminals always begin a connection using the G.711 standard as this allows them to instantaneously communicate with all other H.320 terminals as well as normal telephones. Most systems then attempt to negotiate to move to either G.728 to maximize the channel available for video and data, or G.722 to improve the audio quality.

10.3.3   H.221 and H.242 — the channel aggregation and multiplex standards

H.221 and H.242 perform the following roles.

- H.221 provides facilities to allow multiple 'B' channels to be aggregated together. This allows any differential delay between the two channels in a 2B ISDN call to be removed, creating a single 128-kbit/s channel rather than two 64-kbit/s channels.

- H.242 provides facilities to allow a transmitting terminal to determine the capabilities of the receiving terminal. This allows terminals to automatically disable functions which are not supported by the other terminal. For example, if the far terminal states that it does not support G.728 audio, the sending terminal will not use it during the call.

- H.242 and H.221 provide facilities for defining and signalling the multiplex structure to the remote end so that it is capable of identifying and extracting the individual audio/video/data components. This multiplex structure can be changed every 20 ms during a call, which allows the trade-off between audio, video and data to be dynamically varied to meet the user's requirements.

10.3.4   T.120 — the data standard

T.120 is an umbrella standard for a series of standards which together define the protocols required for collaborative working (see Fig. 10.3). The T.120 standards

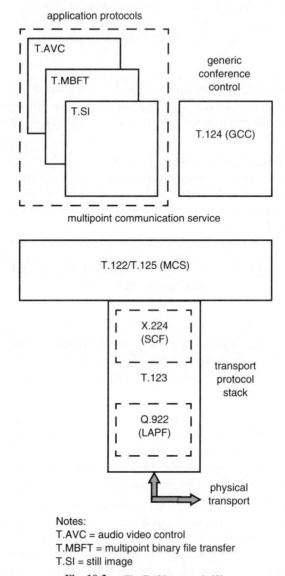

Fig. 10.3   The T.120 protocols [9].

are layered, covering all aspects of the protocol from the transport layer to the conferencing layers and finally the shared application layers. Although T.120 is the last of the multimedia standards to be finalized, it has now attracted a sufficient international following to ensure that it remains the data conferencing standard. Microsoft has recently included T.120 as an add-on to Windows 95 and announced that it will be fully incorporated in the next release.

• The transport stack T.123 [12]

T.120 has been designed to provide multipoint data services across a wide variety of data networks such as PSTN, ISDN, LAN and B-ISDN. The transport protocol stack is the layer where the network dependence is removed. The top of the transport layer provides four prioritized, error-free channels between the two ends. The software above the transport stack can assume an error-free transmission of the data submitted to the channel. The four channels of differing priority are provided to allow real-time data (e.g. cursor movements) to pre-empt background data transfers (e.g. file transfers). The transport protocol stack is changed depending on the target network. For the ISDN the lower levels of the Q.922 protocol result in synchronous (HDLC) packets being presented to the network. For the PSTN the lower levels of the Q.922 protocol implement an asynchronous (RS232-type) protocol.

• The multipoint communications service (T.122 [13]/T.125 [14])

The multipoint communications service (MCS) provides a link layer which allows any terminal within a conference to communicate with any other terminal(s) as required. Each pipe (link) is named and connects a defined set of terminals together, with a defined priority of transport. Hence, if a conference consisted of five terminals A, B, C, D and E, then MCS provides the capabilities for A, B and E to set up a pipe between them for a private, background, file transfer while at the same time operating a pipe between all five terminals for a multipoint shared application. MCS supports up to 64 000 simultaneous connections within a call, which should be enough for any requirement.

Hence, MCS can be viewed as a telephone network which allows point-to-point and multipoint calls between the members of a conference. Members can set up as many different point-to-point and multipoint calls as they require, to allow only some sub-groups within the conference to see some data sets.

• The generic conference control (T.124 [15])

The generic conference control (GCC) module manages the setting up and tearing down of pipes through the MCS. It receives requests for service directly from user applications as well as from the application protocols. The GCC is also responsible for the management of a multipoint conference including chairperson control and reservation functions.

- The application protocols — still image (T.126/T.SI [16])

  The still image application protocol is probably the most important protocol to be developed. It defines the way in which still image and annotation information will be shared within a collaborative environment. T.SI provides protocols to enable still-image transfer (facsimile), shared annotation (whiteboard/chalkboard) and simple remote control applications (application sharing). These protocols allow manufacturers to connect simple and complex applications together, with all applications being able to operate in a compatible manner. While some manufacturers may define extensions to the basic still-image protocol, they will need to provide an automatic fall-back position for standards-compliant communications with other terminals.

- The application protocols — multipoint binary file transfer (T.127/T.MBFT [17])

  The multipoint binary file transfer protocol is the basic file-transfer protocol built into T.120. It includes all the functionality required to operate in a multipoint environment as well as the basic point-to-point transfers. As its name suggests, T.MBFT is designed for the transfer of binary data as well as normal ASCII text. Hence any file on any PC in a conference can be transferred to any number of the other PCs in the call.

- The applications protocols — audio/video control (T.AVC [18])

  The audio/video control protocol provides the facilities to manage audio and video data streams outside T.120. Consequently, in an H.320 terminal, the T.AVC protocol would provide remote control of cameras (pan, zoom, document versus face-to-face), control of the switching within a multipoint control unit (MCU), and any other control facilities required within a multimedia call. T.AVC does not carry any real-time audio or video information. It is purely a control channel to provide sophisticated control for the real-time audio and video streams within a call.

This section has given an overview to the T.120 standard, describing the work of the ITU-T in defining the transmission protocols associated with shared working. While this work is the minimum required to achieve interworking across the world, the computer industry would like further standardization to occur to allow applications from many vendors to run on any T.120 protocol stack. This work requires that an API be defined at each layer in the protocol stack. This API work is being performed by an industry group known as the IMTC [19], which includes representation from PTTs, product manufacturers and, of course, the computer industry. Microsoft, Intel, BT and many others are working together to define the APIs which will allow a chalkboard application from one vendor to work seamlessly with a T.120 protocol stack from another.

10.4   MULTIPARTY CONFERENCING

The discussion so far has concentrated on the design of the terminal equipment required for desktop multimedia. However, another very important area of development is the technology within the network to allow effective group working between desktop multimedia terminals world-wide. This group working is known as 'multiparty conferencing' and a network system that provides support for this type of operation is known as a 'multipoint control unit' (MCU).

The goal of multiparty operation is to allow users with desktop multimedia terminals, designed to the standards described in the previous section, to inter-work together as if they were physically co-located. Existing MCU equipment supports voice bridging and video switching. Future systems will include much more sophisticated services that will be controlled by the T.120 standards described in the previous section.

- Audio within a multiparty call

  The audio technology for multipoint operation is very similar to the technology provided in a traditional audioconferencing bridge. The multipoint unit connects all users together by mixing the audio signals from everyone who speaks and transmitting the mixed signal back to all participants. In a multimedia conference environment the speech paths can also be controlled by the T.120 data path so that one participant (usually the chairperson) can control who may speak. As the technology matures, other control paradigms will be developed to assist in the control of large meetings.

  Looking to the future, it is likely that new standards will allow the transmission of stereo imaging to the participants in a conference, helping to create a feeling of audio presence. Each speaker's voice will be imaged to a virtual location at the receiving terminal, making it easier to identify the speaker in a large conference.

- Video within a multiparty call

  The video technology for multipoint operation is very different to the audio technology. Unlike audio, it is not possible to mix the pictures of different speakers together and produce any meaningful results. Hence the video technology relies on switching the video picture to look at the current speaker. This is the technology that is currently available within videoconference MCUs. The newest MCUs are now beginning to include the chairperson control facilities from T.120, which allow one participant to take control of the conference.

Research into the future of video within multiparty conferences is concentrating on ways to transmit multiple video streams to a single terminal. These multiple streams would allow the viewing of two or more participants at the same time, although the actual quality of each image would depend on the limitations of the channel bandwidth available. As these more sophisticated video systems become available, then it will become necessary to develop mechanisms to control them.

These 'continuous presence' video systems will allow users to monitor the reactions of selected other participants, while still focusing on the speaker.

- Data within a multiparty call

The T.120 suite of standards, described in section 10.3 above, is designed to operate in multipoint as well as point-to-point environments. In fact, the standards include two protocols, T.GCC [15] and T.AVC [18], which are specifically designed to control a multipoint conference. All the collaborative-working applications have been developed with multiparty conferencing in mind. This leads to chalkboard-type applications that can share images across a conference, or application-sharing applications that will allow all the participants of a conference to jointly author a document.

These facilities and others, like the ability to broadcast a file to some or all conference participants, will radically change the way people view electronic meetings.

When the audio, video and data technologies described above are integrated in the next generation MCUs, the user will be presented with a sophisticated, collaborative working environment, which will remove many of the barriers currently imposed by distance. Instead of jumping on to a plane to attend a meeting, the user will be able to dial up access, see all the participants and jointly work with them in a virtual electronic meeting environment. Initially systems will struggle to provide a scenario as good as an existing meeting, but, as users become more familiar with the tools and the systems become more sophisticated, the environments will expand to provide the facilities of the best designed meeting rooms directly to a user's desktop.

10.5   DESKTOP MULTIMEDIA TERMINALS IN THE REAL WORLD

The previous sections of this chapter have explored the make-up of a desktop multimedia system. This section of the chapter will look at how one of these terminals operates in a world environment, which is populated with terminals of many different types and from different manufacturers.

Initial trials with videophones and desktop videophones have highlighted a number of 'telephony'-related requirements for this type of equipment. In particular, while the manufacturers of these terminals have tended to think of them as sophisticated meeting aids, the customers view them as sophisticated telephones. This implies that the desktop terminal needs to behave like a telephone, as well as providing the functionality of a multimedia terminal. How do telephones behave? Some of the features which most managers take for granted are:

- the ability to route all incoming calls via a secretary;

- having only one phone on the desk — if it is in use then incoming calls are routed to the secretary.

These points are very important if users are going to migrate from existing telephones to using desktop multimedia systems. If users keep their standard phones and use desktop systems as well, then the latter will quickly be relegated to second place and will only to be used for previously planned, structured meetings, missing much of its potential use.

So it would seem that for a desktop system to be truly successful it needs to meet the telephony needs of the end user as well as the multimedia needs. However, the structure of the products and standards described in the previous sections have been designed with these needs in mind.

10.5.1   Simple telephony

To provide a basic telephony service is actually quite a challenging requirement for a multimedia system. Throughout Europe the basic telephony service is regulated to ensure that all terminals meet the performance requirements defined in the NET33 [20] standard. If a user's phone is to be replaced with a desktop system, then the desktop system will also need to conform to the requirements of NET33. However, users demand more from their basic phone than they do from their PC. They do not expect their PC to work until they have switched it on, whereas the phone is expected to work all the time. Similarly, they expect their PCs to crash occasionally, or even often, whereas they do not expect their phone to crash. These requirements lead to the conclusion that the PC and the phone should be separated, which implies that the desktop multimedia system should be hosted from a telephone-like instrument rather than trying to replace it.

10.5.2   PABX functionality

To mount a desktop multimedia product behind a PABX requires that the desktop product is capable of all the PABX functions available on a normal telephone. This can put considerable demands on the desktop product, particularly if these PABX functions are also expected to work with multimedia calls. However, the desktop system also has advantages over the simple phone, the most obvious being that it has the PC display available to represent some of the more complex PABX facilities. As desktop systems become fully integrated into the telephony environment, users are likely to see becoming available PABX/telephony interfaces that are much more user friendly, with the simple facilities available via both PC and telephone handset interfaces and the more complex facilities available via the PC interface only.

10.5.3   Interworking

As far as the user is concerned, the desktop multimedia terminal is just a phone with more functionality. If the user calls a normal phone, mobile phone or any other existing telephony service, the desktop system is expected to connect directly and behave just like an ordinary phone. However, if the user calls an endpoint which is capable of multimedia functionality, then the user expects the terminal to automatically enable as much functionality as is available to both ends. This process of starting at the lowest level functionality (simple telephony) and building the call up to the highest level supported by both ends is automatically carried out using the standards H.221 and H.242. As time progresses and more standards become available for multimedia over the PSTN, ISDN and other networks, terminals will have to become more sophisticated to maintain this total interworking capability. Without full interworking, users will quickly get very frustrated as calls start failing for unexplained reasons. Interworking remains one of the major challenges of the multimedia age.

10.6   CONCLUSIONS

Looking to the future, it is clear that desktop multimedia is going to dramatically change our lives. This chapter has looked at the technology behind the desktop multimedia equipment which is currently being deployed in the market-place and discussed some of the issues which are still being addressed. It has shown that ISDN-based products are available now that can significantly improve the communications within many companies, leading to enhanced team working and major productivity increases.

Over the next few years, products will become available which operate over the existing LAN-based networks and the existing PSTN telephone network. The performance of these systems is still to be proven and while there is little doubt that PSTN-based products will become very popular, the telephony performance of a LAN-based product is still to be proven. The success or otherwise of all of these products will ultimately depend on the ease of use and functionality provided.

As time progresses, the number of variants of a terminal giving different price/performance trade-offs will increase and the challenge for the industry will be to contain these systems to ensure continued interoperability. In the next decade there will undoubtedly be an explosion in the penetration of multimedia services, leading to radical changes in the way that people do business.

APPENDIX

Glossary

API

Application programmers interface — a software interface inside the computer, which allows software from different suppliers to interwork.

ITU-T

International Telecommunication Union — a Geneva-based organization which brings the world's PTTs and other interested parties together to define the standards for telecommunications around the world (previously known as the CCITT).

Desktop multimedia

Multimedia communications integrated into a user's desktop terminal (PC or workstation) to allow full video-telephony and multimedia-communications functionality from the single terminal.

G.711 [4]

The international (ITU-T) standard for telephony quality audio coding. This standard supports two incompatible modes, 'A'-Law which is used throughout Europe and 'μ'-Law which is used in the US and Japan (audio bandwidth — 3.4 kHz, digital data rate: 64 kbit/s, 56 kbit/s).

G.722 [6]

The international (ITU-T) standard for high-quality audio coding. This standard supports medium-wave radio quality audio (audio bandwidth — 7 kHz, digital data rate — 64 kbit/s, 56 kbit/s and 48 kbit/s).

G.728 [5]

The international (ITU-T) standard for low bit rate audio coding. This standard supports G.711 quality audio at a quarter of the data rate (audio bandwidth — 3.4 kHz, digital data rate — 16 kbit/s).

H.221 [7]

The international (ITU-T) standard for the aggregation of multiple 64 kbit/s channels together, and the multiplexing of audio, video and data on to the aggregated channel.

H.242 [8]

The international (ITU-T) standard for the negotiation of capabilities between terminals to ensure that one terminal does not try to perform an operation which is outside the capability of the other terminal.

H.261 [3]

The international (ITU-T) standard for video coding within a conversational audiovisual call.

H.320 [10]

The international (ITU-T) umbrella standard which covers multimedia communications across the ISDN. This standard calls up the other G, H, I and T series standards listed here.

I.420 [2]

The international (ITU-T) standard for ISDN BRI.

ISDN BRI

ISDN basic rate interface — the standard digital telephone interface used throughout the world, conforming to I.420. This interface presents two 64-kbit/s bearer 'B' channels and one 16-kbit/s data 'D' channel. The 'B' channels are used to connect together the parties in a call and provide either 64 kbit/s or 128 kbit/s of channel capacity between the desktop multimedia systems. The 'D' channel is used to exchange signalling information with the telephone network for call set-up, etc.

Multimedia

Computer industry term to describe the merging of audio and video with data. The audio and video involved with standard multimedia applications is usually stored locally on the PC or workstation and played back at the same time as the associated data is accessed.

Multimedia communications

Communication industry term to describe the merging of real-time audio and video with all forms of data. The audio and video involved in multimedia communications applications are sourced in real time from the remote party(ies) in the call. The data element can include any data that is resident on a PC, workstation or other digital device (e.g. Group 4 Facsimile).

PC

Personal computer — this term has been used in this chapter to include IBM compatible PCs, Apple Macintosh PCs and workstations of all types. It covers any computing system that might reside on a user's desk.

PSU

Power supply unit.

T.120 [9]

An umbrella standard for a series of standards which define the protocols for collaborative data working between two or more computers in a conference.

Videophone

A terminal whose major purpose is to provide real-time audio and video functionality. Videophones often include a data port which allows them to be connected to a PC or workstation to provide multimedia communications functionality as well.

REFERENCES

1. Allen S and Wolkowitz C: 'Homeworking: Myths and Realities', Macmillan Education Ltd (1987).

2. ITU-T Recommendation I.420: 'Basic user-network interface', [Blue Book Fasc III.8] (1988).

3. ITU-T Recommendation H.261: 'Video codec for audiovisual services at p × 64 kbit/s', [Rev. 2] (1993).

4. CCITT Recommendation G.711: 'Pulse code modulation (PCM) of voice frequencies', [Blue Book Fasc III.4] (1988).

5. ITU-T Recommendation G.728: 'Coding of speech at 16 kbit/s using low-delay code excited linear prediction', [New] (1992).

6. CCITT Recommendation G.722: '7 kHz audio coding within 64 kbit/s', [Blue Book Fasc III.4] (1988).

7. ITU-T Recommendation H.221: 'Frame structure for a 64 to 1920 kbit/s channel in audiovisual teleservices', [Rev. 2] (1993).

8. ITU-T Recommendation H.242: 'System for establishing communication between audiovisual terminals using digital channels up to 2 Mbit/s', [Rev. 1] (1993).

9. ITU-T Recommendation T.120: 'Transmission protocols for multimedia data conferencing', (Draft) (1995).

10. ITU-T Recommendation H.320: 'Narrow-band visual telephone systems and terminal equipment', [Rev 1] (1993).

11. ITU-T Recommendation H.324: 'Visual-telephone terminals over GSTN', (1996).

12. ITU-T Recommendation T.123: 'Protocol stack for audiographics and audiovisual teleconference applications', (1993).

13. ITU-T Recommendation T.122: 'Multipoint communication service for audiographic and audiovisual conferencing', (1993).

14. ITU-T Recommendation T.125: 'Multipoint communication service protocol specification', (1994).

15. ITU-T Recommendation T.124: 'Generic conference control for audiovisual and audiographic terminals', [rev 2] (1996).

16. ITU-T Recommendation T.126: 'Still image multipoint and annotation protocol specification', (1996).

17. ITU-T Recommendation T.127: 'Multipoint binary file transfer protocol specification', (Draft) (1995).

18. ITU-T Recommendation T.130: 'Real-time architecture for multimedia, conferencing', [Determined] (1996).

19. IMTC url:http://www.csn.net/imtc/

20. NET33 (ETS 300 085): 'Integrated services digital network (ISDN) — 3.1 kHz telephony service: attachment requirements for handset terminals', (1993).

11

MULTIMEDIA IN INTER-CONNECTED LAN SYSTEMS

W Bunn

11.1　INTRODUCTION

Networked multimedia applications can be delivered to the desk today via PBXs or LANs, but both systems have their limitations. PBX solutions can provide low end-to-end delay, but only at relatively low bandwidth, whereas LANs can deliver high bandwidth, but the delay is unpredictable. The importance of these limitations depends on the relative significance of the real-time (audio and video) and non-real-time (data) components that are used by the multimedia applications. This chapter examines the delays that can be encountered in interconnected LAN systems and possible ways of constraining them.

11.2　MULTIMEDIA APPLICATION CATEGORIES

Multimedia applications can be divided into two broad categories, namely 'interactive' and 'retrieval'. Both types can include any mix of audio, video and data, but the interactive category is much more sensitive to delay and delay variations (jitter). Video telephony clearly comes in the interactive category and human factors dictate that the end-to-end delay must be kept to a minimum. Other interactive applications that are sometimes associated with video telephony are shared word processors, spreadsheets or drawing packages.

Retrieval applications such as distance learning are more delay-tolerant, since users will readily accept a delay of a few seconds between requesting information and having the information presented on their screens. Retrieval applications can also benefit from the use of asymmetric encode/decode algorithms which use a simple, low-delay, decode algorithm at the expense of a more complex encoding algorithm. This is acceptable for retrieval applications since the encoding is only done once and does not have to be done in real time.

Delay variation is more problematical than absolute delay for both categories of applications, but here again retrieval applications have an advantage since buffers can be introduced to smooth out any variations and thereby minimize the impact on the end user. The following discussion applies equally to all types of applications, but the use of retrieval applications may grow faster in the short term because they are less demanding on the technology.

11.3   LOCAL AREA NETWORKS

Traditional LANs such as Ethernet™ were originally designed to interconnect a number of computers and peripherals across a shared, passive co-axial bus. Attached nodes check that no other nodes are using the bus before they transmit each packet of data. This access protocol can obviously introduce significant delay variations dependent entirely on the volume of traffic from other nodes attached to the network. The problem can be constrained by limiting the number of nodes connected to each network and today, with structured wiring where the co-axial bus is replaced with twisted pairs, it is common to have one node per network.

The hubs used to interconnect such networks do so with multiport bridges and/or routers [1] which use store-and-forward technology. For data applications this is fine, since it enables a high level of error detection and correction (by re-transmission) to be introduced, but the delay variations that can also be introduced are likely to cause havoc to any real-time elements of multimedia applications. In the short term, a number of users will tolerate the delay problems, especially on lightly loaded networks. However, most users are likely to wait for the next generation of routers that will incorporate the new IETF standards which specifically address this problem. The first such routers are expected to become available in late 1996.

The following description, together with Fig. 11.1, illustrates how the main components of a LAN-based network contribute to the overall end-to-end delay as seen by an interactive application. Additional unlabelled boxes are shown in Fig. 11.1 to emphasize the fact that LANs are shared. These boxes could be personal computers, workstations, print servers, or bridges and routers linking to other networks.

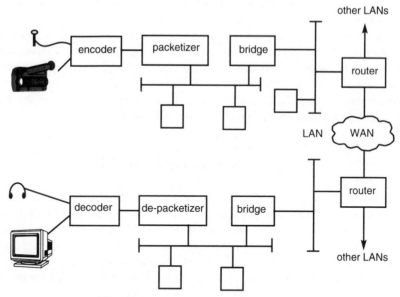

Fig. 11.1   Typical extended local area network.

The encoder converts the analogue signal into a digital data stream. The delay is dependent on the coding scheme and compression algorithm used. Algorithms that provide high compression ratios also introduce higher delays. For H.261-encoded video at 128 kbit/s the delay is approximately 180 ms. The delay through the packetizer depends on the packet size and for Ethernet systems this can vary from 64 bytes to 1518 bytes (including overheads), with corresponding delays of between 4 ms and 95 ms. This presents a classic 'catch 22' problem in that to keep the end-to-end delay down, small packets should be used, but due to the fixed per-packet overheads small packets increase the overall level of network traffic. This delay is entirely deterministic and Table 11.1 illustrates how the delay varies with packet size and clock rate. The delay across the physical LAN is measured in microseconds and can generally be ignored.

Bridges typically extend LANs within a building or campus site and although Fig. 11.1 shows a two-port bridge it is common to have multiport units capable of switching the traffic between any ports. Bridges read in the complete data packet and check it for errors before forwarding it. Assuming the data is error free, the bridge accesses the destination network in exactly the same way as any other device and must wait until the network is idle before transmitting the data. Data that cannot be immediately transmitted is temporarily stored in a FIFO (first-in-first-out) store and this is where the delays can build up depending on the level of network traffic. Under worst-case conditions bridges can introduce delay variations of a few hundred milliseconds although a few tens of milliseconds is more typical.

Table 11.1   Serialization delay (ms).

Clock rate	Packet size (bytes)			
	64	256	1000	1518
64 kbit/s	8	32	125	190
128 kbit/s	4	16	63	95
384 kbit/s	1	5	21	32
1500 kbit/s	0.3	1	5	8
2048 kbit/s	0.3	1	4	6

Routers are also multiport devices and are typically used to interconnect LANs across wide area networks although they can be used in place of bridges to give more control over the traffic flows. Routers operate in a similar manner to bridges, but the delay variations can be much more significant, particularly when used to connect across wide area networks [2]. This is due to the fact that the wide area network bandwidth is typically only a fraction of the LAN bandwidth and the delay through the FIFO varies as the traffic bursts come and go. Table 11.2 illustrates the delay that can build up in a router depending on the size of the FIFO and the line rate of the wide-area network interface.

Table 11.2   Buffer capacity (seconds).

Clock rate	Memory size (kbytes)		
	128	512	1024
64 kbit/s	16.0	64.0	128.0
128 kbit/s	8.0	32.0	64.0
384 kbit/s	2.7	10.7	21.3
1500 kbit/s	0.7	2.7	5.5
2048 kbit/s	0.5	2.0	4.0

The delay across the wide area network is approximately 8 μs/mile; therefore if the link is short this delay can be ignored, but not if the link goes around the world or involves a satellite hop.

The depacketizer introduces similar delay to the packetizer and the video decoder introduces approximately 120 ms assuming 128-kbit/s H.261 video.

It can be seen from the above figures that routers and to a lesser extent, bridges are the main source of delay variation and hence represent the key problem areas in supporting interactive multimedia applications on distributed LANs.

Figure 11.1 shows a single bridge and router at each end of the extended network, but real systems are more likely to have a number of these units in series and hence the delay variation problems become even worse.

Traffic profiles on today's LANs vary widely but most users would not notice if a few multimedia demonstrators were added to existing networks. However, large-scale roll-out of such applications would require network upgrades. LAN vendors are coming up with a number of schemes to reduce the problems and extend the life of the installed systems, but to date such solutions only work in specific cases and are proprietary in nature. A new standard that directly addresses the delay problem is expected to be ratified during 1996. This standard introduces new, quality-of-service information into the packet headers so that the intermediate systems (routers) can identify the different types of data and expedite the forwarding of real-time data at the expense of other traffic.

Another solution is isochronous Ethernet (IEEE 802.9a). This multiplexes packet and circuit-switched traffic over the same physical wire. Today this offers the best technical solution for interactive multimedia applications, since the real-time interactive traffic can go over the low-delay connection-oriented network and the data traffic over the connectionless network. Some switch vendors are adding this as an optional interface card in their PBXs but it remains to be seen whether the market takes this up.

11.4   CONCLUSIONS

If the packet network world gets its new resource reservation standard correct, then future multimedia applications that include audio and video will be delivered to the desk via LANs. PBXs will remain dominant for low-bandwidth applications such as voice only, but there is little if any growth in this area and the LAN could displace rather than work alongside the PBX in many businesses.

REFERENCES

1.   Pearlman R: 'Interconnections (Bridges and Routers)', Addison-Wesley (1992).

2.   Miller M: 'Internetworking (LAN to LAN, LAN to WAN)', Prentice Hall (1991).

12

A REVIEW OF EYE-TO-EYE VIDEOCONFERENCING TECHNIQUES

D A D Rose and P M Clarke

12.1 INTRODUCTION

In an age of increased communication, with greater availability and use of broadband telecommunications networks, there has been a corresponding expansion in the use of video-based telecommunications.

As a result, video-codec equipment is becoming more widely available, at a lower price. One key application of video-based telecommunications is video-conferencing. Videoconferencing enables people who are remotely located from one another to communicate using both sight and sound. In effect, videoconferencing reduces the communications void that exists between traditional audio-based telephone conversations and real-life person-to-person meetings. Figure 12.1 illustrates a typical videoconferencing display and camera configuration.

Although videoconferencing systems are very effective as a means of remote person-to-person communication, they do not totally fill the need for actual person-to-person meetings, with associated time and travelling costs. In the UK alone, it is estimated that £14 billion is spent on business travel per year.

Several innovative approaches have been described that increase the feeling of presence between participants in a videoconference meeting, in effect increasing the level of 'telepresence'. Telepresence may be considered as the use of technology to establish a sense of shared presence or shared space among geographically separated members of a group. The advancement of telepresence generally requires the integration of telecommunications, computing and audio/video technologies.

One method of increasing the feeling of telepresence in videoconference meetings is to use large screen displays. This results in a more realistic and life-size image of the remote participant, with more easily recognizable body language.

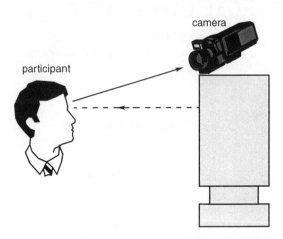

Fig. 12.1 Conventional videoconferencing arrangement.

One problem that reduces the level of telepresence during videoconference meetings is the inability of participants to maintain eye contact with each other, with a corresponding loss of rapport. As conventional videoconferencing systems typically place the camera either to one side or above the display screen, the camera sees an off-angle view of the participant and eye contact between participants cannot be achieved. This unwanted effect is further worsened by the increased use of large screen displays, as might be used in the studio.

This chapter provides the reader with a basic understanding of eye contact and gives a review of various proposed solutions. It also describes some of the experimental work being carried out at BT Laboratories.

12.2   WHAT IS EYE CONTACT?

Eye contact in this context refers to the sensation of looking into another person's eyes.

The perception of eye contact is very subjective, varying from absolute line of sight between eyes to generally looking in the direction of a person. Therefore it is difficult to measure eye contact in the laboratory [1]. Psychologists often refer to gaze (looking directly at the face of another person) and mutual gaze (two people looking directly at each other), though these are used interchangeably with eye contact in general usage.

The role of eye contact in communications is more important than the definition. Eye contact performs four key functions [2-4]:

- regulating the flow of conversation — eye contact is used to regulate invitations to start a conversation and cues to speak (interestingly these cues can be interpreted differently by people of different ethnic origin, interrupting the flow of conversation [5]);

- monitoring feedback — the presence of gaze can convey interest, attention and attraction, and likewise lack of gaze or breaking off gaze can convey the opposite (this feedback can unintentionally increase attraction of males to females, though females seem less affected [6, 7]);

- expressing emotion — along with the brows and mouth, the eyes are one of three main elements that convey facial emotion; people have been described as having serious or smiling eyes, and obscuring the eyes, for instance behind sunglasses, causes tension for the other person [8];

- communicating the nature of the interpersonal relationship — gaze can be used to convey like and dislike, and an intensity of positive or negative feeling; it can also convey rank or status between people — a higher status person often displays less eye contact than their lower status converser, implying less need to persuade [9, 10].

Non-verbal contact contributes much to an interaction — we retain as much as 83% of what we see, yet only 11% of what we hear (according to Karch). How much eye contact contributes is not known, but it is certainly significant [11].

12.3   EYE CONTACT SOLUTIONS

Various methods have been proposed that are intended to increase the feeling of eye contact between participants during videoconference meetings. These methods may be categorized into the following groups:

- light division, where the camera and display simultaneously share the same optical path;

- space division, where the camera may occupy part of the display area, such as using a pin-hole arrangement;

- time division, where the display screen material is switched rapidly from an optically opaque and translucent state in synchronism with the display and camera equipment;

- image processing, where the facial features that affect eye contact are artificially superimposed to give the impression of eye contact.

12.3.1   Light division

12.3.1.1   Half-transparent mirror arrangement

This arrangement is possibly the most conventional way of effecting eye-to-eye contact, one of which is available as a commercial product [12]. Light division is used to achieve the effect of eye contact by arranging the camera so that it simultaneously shares the same light path as the display screen. A half-transparent mirror (HM) is placed in front of the viewed screen at an approximate angle of 45°, as illustrated in Fig. 12.2.

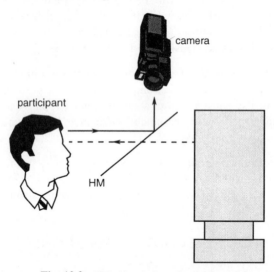

Fig. 12.2   HM videoconferencing configuration.

The HM has the effect of optically changing the axis of the camera. By positioning the camera as shown, the image seen by the camera is the reflected, face-on view of the participant. In practice, a second mirror can be located in the optical path of the camera in order to negate the effects of the single reflection image reversal.

12.3.1.2   Blazed half-transparent mirror

A development of the HM arrangement, referred to as the blazed half-transparent mirror (BHM) has been described and analysed elsewhere [13, 14]. The name

BHM arises due to the structural similarity with a blazed grating. This technique is based on a novel HM arrangement that can be placed in parallel to the display screen surface. This has the advantage that, while still enabling eye contact, the BHM material does not protrude forward of the display as in the HM arrangement. This enables design of a more compact and less intrusive videoconferencing terminal that enables participants to effectively touch the display surface.

The formation of the novel BHM structure can be considered in stages as illustrated in Fig. 12.3. A conventional HM surface, normally inclined at an angle of 45°, is divided into millimetre-sized pieces, referred to as micro-HMs. By moulding many micro-HMs on to a single sheet, a BHM surface is formed.

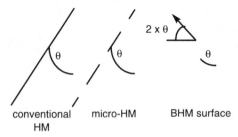

conventional   micro-HM   BHM surface
HM

Fig. 12.3   Formation of BHM sheet.

In practice the BHM sheet would be used as a conventional HM but vertically oriented rather than inclined at 45°, as illustrated in Fig. 12.4.

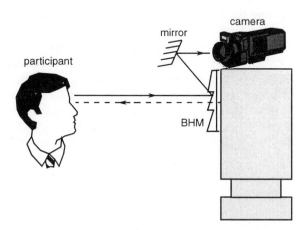

Fig. 12.4   BHM videoconferencing configuration.

12.3.2   Space division

The space-division method for achieving eye contact between the camera and participant uses a miniature camera. The camera is located behind the display surface, at a height corresponding to the participant's eye level, as illustrated in Fig. 12.5.

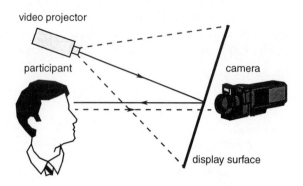

Fig. 12.5   Space-division videoconferencing configuration.

A small hole, the size of the camera lens aperture, is placed in the screen surface to give the camera, located behind the screen, a clear view of the face-on image of the participant. The viewed image is generated by a front-projection video projector. Using this arrangement, some form of control is required to prevent the camera seeing the light output from the video projector. This might be achieved using either a camera arranged such that its field of view does not include the projector, or possibly by using polarized material between the projector and camera, or alternatively using some form of time division control.

A similar system has been described [15], consisting of an overhead video projector, display screen and two miniature cameras mounted behind the screen. A Macintosh Quadra computer is also used to generate graphics information using the data compatible video projector. An additional feature of this system is the use of two directional microphones, placed either side of the screen. Should there be multiple participants or a single participant located off-centre to the screen, the microphones detect the direction of the speech and hence automatically select the most appropriate camera.

12.3.3   Time division

The time-division method of achieving eye contact is obtained by arranging the camera to time-share the same optical path as that which the participant uses to view the display screen. This can be achieved by placing the camera and a rear-projection video projector behind a switchable liquid-crystal display (LCD) panel, as illustrated in Fig. 12.6.

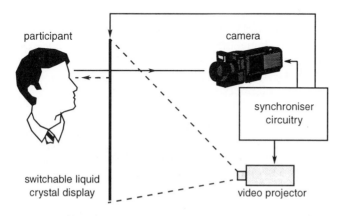

Fig. 12.6   Time-division videoconferencing configuration.

The LCD panel is switched repetitively from an opaque to translucent state in synchronism with the camera and video projector. The camera and video projector are switched alternatively such that when the LCD panel is opaque the video projector is active, and when the LCD panel is translucent the camera is active.

This system has an advantage that there is no obtrusive HM or BHM surface placed in front of the display screen, making it especially practical for larger display screen applications. However, the camera and video projector are only active for half of the total time. This results in a 3-dB reduction in the average light level viewed by the camera and similarly a 3-dB reduction in the display illumination intensity. The technique has been described in detail [16-18] using a 40-inch switchable LCD panel.

12.4   CONCLUSIONS

This review has concentrated only on methods of achieving eye contact in videoconference meetings. It has not considered other techniques for increasing the effect of telepresence. These might include the ability to maintain eye contact between multiple participants [19, 20], provision of three-dimensional imaging effects [21-25], or the sharing of interactive workspace environments [26-28].

All of the reviewed eye contact methods achieve eye contact without the need for the participants to wear special equipment.

The use of eye contact in videoconference meetings enables the participants to communicate more naturally, even though they are in separate locations.

The benefits of using such systems are the reduction in time and costs that would otherwise be incurred travelling to meetings by car, train or plane.

Several videoconferencing terminals that achieve eye contact between participants have been constructed at BT Laboratories. The methods used have been based on variations of the HM technique, with added features including a shared interactive workspace. Initial user feedback has been very positive. Other variations are being considered that combine time and space division techniques, using micro-louvred sheets and linear polarized materials.

REFERENCES

1.   von Cranach M and Ellgring J H: 'Problems in the recognition of gaze direction', in von Cranach M and Vine I (Eds): 'Social communication and movement', Academic Press, New York (1973).

2.   Argyle M, Ingham R, Alkema F and McCallin M: 'The different functions of gaze', Semiotica, 7, pp 19-32 (1973).

3.   Kendon A: 'Some functions of gaze-direction in social interaction', Acta Psychologica, 26, pp 22-63 (1967).

4.   Knapp M L: 'Non-verbal communication in human interaction' (2nd ed), Holt, Rinehart & Winston, New York (1978).

5.   LaFrance M and Mayo C: 'Racial differences in gaze behaviour during conversations: two systematic observational studies', Journal of Personality and Social Psychology, 33, pp 547-552 (1976).

6.   Knackstedt G and Kleinke C L: 'Eye contact, gender, and personality judgements', Journal of Social Psychology, 131, No 2, pp 303-304 (1991).

7.   Kleinke C L, Bustos A A, Meeker F B and Staneski R A: 'Effects of self-attributed and other-attributed gaze on interpersonal evaluations between males and females', Journal of Experimental Social Psychology, 9, pp 154-163 (1973).

8.   Yoshida F: 'An analysis of visual behaviour in conversational dyads', Japanese Journal of Experimental Social Psychology, 20, No 2, pp 109-118 (1981).

9.   Exline R V: 'Visual interaction: the glances of power and preference', in Cole J K (Ed): 'Nebraska Symposium on Motivation', 19, University of Nebraska Press, Lincoln (1971).

10. Exline R V, Ellyson S L and Long B: 'Visual behaviour as an aspect of power role relationships', in Pliner P, Krames L, and Alloway T (Eds): 'Non-verbal communication of aggression' (2nd ed), Plenum Press, New York (1975).

11. Droney M and Brooks C I: 'Attributions of self-esteem as a function of duration of eye contact', Journal of Social Psychology, 133, No 5, pp 715—722 (1993).

12. Alcatel/Standard Eletrik Lorenz AG: 'The future of the telephone is looking good, ISDN videophone'.

13. Arai H, Kuriki H and Sakai S: 'New eye-contact technique for video phones', SID International Symposium Digest of Technical Papers, XXIII, pp149-152 (1992).

14. Kuriki M, Arai H, Arai K and Sakai S: 'Resolution analysis of eye contact technique using BHM', Japan Display 92, Proceedings of the 12th International Display Research Conference, pp 399-402 (1992).

15. Global Talk Power, German article describing the 'talk tower' project (1991).

16. Shiwa S and Ishibashi M: 'A large-screen visual telecommunications device enabling eye contact', SID International Symposium Digest of Technical Papers XXII, pp 327-328 (1991).

17. Shiwa S, Nakazawa K, Kamatsu T and Ichinose S: 'Eye-contact display technologies for visual telecommunications', NTT Review, 5, No 2, pp 67-73 (March 1993).

18. Shiwa S: 'A large-screen visual telecommunications device using a liquid crystal screen to provide eye contact', NTT Human Interface Laboratories, Yokosuka, Tokyo, Japan, presented at SID 91 (1991).

19. Nakazawa K, Shiwa S and Ichinose S: 'Private display method for teleconferences', Japan Display 92, Proceedings of the 12th International Display Research Conference, pp 395-398 (1992).

20. Nakazawa K, Shiwa S and Ichinose S: 'Private-type display for video phones', SID International Symposium Digest of Technical Papers, XXIII, pp 153-156 (1992).

21. Session 9, 3-D Displays, Japan Display 92, Proceedings of the 12th International Display Research Conference, pp 295-322 (1992).

22. Session 38, 3-D Displays, SID International Symposium Digest of Technical Papers, XXII, pp 819-847 (1991).

23. Session 43, 3-D Displays, SID International Symposium Digest of Technical Papers, <u>XXIII</u>, pp 829-848 (1992).

24. Arthur C: 'Get Your Head Around 3-D', New Scientist, p 22 (4 February 1995).

25. Coulter M: 'Enhanced depth on suspended images', Electronic Times, p 16 (17 November 1994).

26. Ishii H and Kobayashi M: 'ClearBoard — a seamless medium for shared drawing and conversation with eye contact', Proceedings of ACM SIGCHI Conference on Human Factors in Computing Systems, pp 525-532 (1992).

27. 'Talking Conference On Board', report describing a large-screen document conferencing system, LiveBoard, launched in the UK by Xerox (1994).

28. Tang J and Minnerman S L: 'Video white board: video shadows to support remote collaboration', Proceedings of ACM SIGCHI Conference on Human Factors in Computing Systems, pp 315-322 (1991).

13

VIRTUAL REALITY — THE NEW MEDIA?

A S Rogers

13.1   BEYOND MULTIMEDIA

When multimedia systems started appearing there were many people who could not see what advantage it would be to be able to play sound from a PC, display high-quality pictures on the screen or see a movie in a window. However, as illustrated throughout the other chapters in this book, multimedia is growing in importance and many multimedia applications are emerging. Three main areas for multimedia applications can be identified. One area is concerned with finding information and is exemplified by multimedia encyclopaedias, sales catalogues, image libraries, etc. Another area is the presentation of information in a structured way for training, sales promotions, etc. The third area, which is now beginning to make an impact on the market, is desktop conferencing which enables people to interact with each other using video as well as voice.

The use of multimedia, whether it includes sound, still image or moving pictures, can convey a very rich message to the user. However, in our home and office environments we not only make use of a variety of '2-D' media but also work in three dimensions. We spread things around so that the things we are currently using are close to us while other things we may need later are further away and less intrusive, but still reachable when required. In a multimedia system, all information is presented to us in 2-D windows, overlapping on a 2-D screen. This screen can quickly become very cluttered and it is often not as convenient to use as a set of books and papers on our desks. Similarly if we are in a room with a large number of people we will be aware of people in different parts of the room, but if we want to speak to someone in particular we move closer to them to catch their eye when it is convenient. Desktop conferencing systems provide the opportunity to see as well as speak to other people and share information with them, but it is still very much a flat world with the picture of the other person fixed in a given place on a 2-D screen.

However, as computing power continues to grow, it is now becoming possible to present information, not just in a 2-D multimedia world, but also in a 3-D virtual-reality world. This enables information to be presented in a much more natural way and opens up many opportunities to improve the ways that users can interact with information and with other people via their computers. This chapter describes some of the ways in which a project called Virtuosi, led by BT, is developing virtual reality and 3-D visualization techniques to support people in their day-to-day work and communications.

13.2   VIRTUOSI PROJECT BACKGROUND

Virtuosi [1] is a collaborative project involving a number of industrial and academic partners. The project is led by BT and the other industrial partners are GPT Limited with GMMT Hirst division (formerly Hirst Research Centre) as a sub-contractor, BICC plc and Division Ltd. The academic partners are the Universities of Nottingham, Lancaster, and Manchester. Also associated with the project are Nottinghamshire County Council and Nottingham Trent University supporting operations in the fashion industry. The project is part of a government programme in Computer Supported Co-operative Work (CSCW) which is jointly funded by the Department of Trade and Industry (DTI), the Engineering and Physical Sciences Research Council (EPSRC) and the industrial collaborators.

The sub-title for the project is 'support for virtual organizations'. In this context, the term virtual organization refers to a group of people who, although formally part of a number of different companies or organizations within the same company have come to work together with a particular set of shared objectives. The formation of the virtual organization might be explicit, for instance a project team within a company, or it may be more informal in the form of some sort of special interest group. Good communications are vital for a virtual organization to achieve its goals. It might seem ideal for everyone to work together in the same office. However, this may not be practical or even desirable since relocation can bring its own problems and disruption. Some form of electronic support is required that can give all the benefits of sharing the same open-plan office, but hopefully without some of the disadvantages. Virtuosi is investigating how such support could be provided in two main application areas. One is in an industrial manufacturing setting where the aim is to form the technical experts and the staff in a number of different factories into a much closer, collaborating community. The other is in the fashion and clothing industry in which designers, buyers, manufacturers and many other specialists are all involved in the complex and time-critical job of creating and selling a garment.

13.3   VIRTUAL REALITY TO SUPPORT GROUPWORKING

Multimedia is already being used to support group working. The most obvious application of this is desktop videoconferencing. This provides a voice and video connection between two users and also enables them to transfer files and share applications. However, full support of groupworking within a virtual organization requires much more than just being able to set up point-to-point communications. It involves the creation of a complete environment that will enable users to interact intuitively and to feel as though they are sharing an office with the other members of the team. There are three main functions that cover most of these requirements. These are really extensions of three main types of multimedia application already identified — finding and viewing information, presenting and interpreting information and interacting with people.

13.3.1   Finding and viewing information

13.3.1.1   The factory pilot

Many functions within a business environment are concerned with locating and retrieving information. As the amount of information grows this can prove to be a difficult task especially when the information is generated and shared by a large number of people. In the factory pilot a study is being made of the operations of a technical centre which provides support for a number of factories across the world. An important resource generated by members of the technical team is the number of papers and reports that document the various studies and investigations in which they are involved. The operations of the technical centre have been studied by sociologists from Lancaster University using ethnography, a technique which concentrates on observing the interactions between people and their environment. The studies at the technical centre have shown the particular importance of documents in the work of the centre, not only as a means of communciation but also for structuring the work they do.

The documents produced by the centre are now being scanned and stored in a database to make them more accessible both to people inside the centre and to factories around the world. The initial implementation of the documentation service will provide a standard set of indexing facilities using keywords and text searching of abstracts, etc. However, studies are under way in Virtuosi to investigate other ways of finding and viewing the reports. One method being studied is called populated information terrains (PITs) (see Fig. 13.1). This uses 3-D to represent different features or parameters of the documents. By choosing suitable parameters for the axes, documents of a similar type will be seen to be clustered together. Colour and shape can also be used to enable the user to see general characteristics of a document at a glance. PITs have another interesting feature in that they not only represent the information in the database in a 3-D manner, but they also provide indications about other people accessing the data. A normal database does not give any indication about other users of the system. Indeed the environment is deliberately designed so that each user is entirely independent and unaware of other users. However, it may be helpful for a user to know if anyone is accessing the same report or cluster of reports at the same time since they may be able to help with a particular problem. Similarly a history of previous accesses may indicate other users with similar interests.

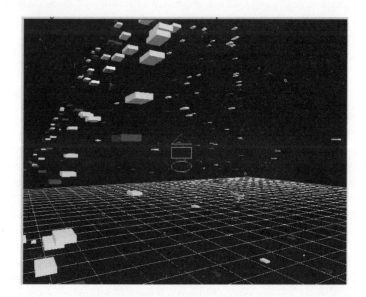

Fig. 13.1   A populated information terrain.

This link between information and people is important. Moreover, in many cases, documents are used not so much as a source of information in their own right as links to people who can help. Indeed, finding people within a real or virtual organization is as important, if not more so, than finding information. It is also a very dynamic issue since the availability of people is continually changing and a user may wish to communicate with different people depending on who is available.

If a group of people are working together in the same office, it is very easy to see who is available. People coming into the office may approach a person they know can help them or someone close by if the first person is not there. The factory pilot is looking at ways to provide this functionality in a virtual environment. An early prototype that has been developed is a visualization of one of the offices within BICC. It is not a fully accurate representation but it gives a good impression of the layout of the office. Each desk is labelled with the name of the person who usually sits there. This world can be 'visited' by a user who can move about freely in the virtual space. It is a shared environment, so more than one user can be present in the virtual space and they are represented by very angular shapes which can at least show where they are and in what direction they are looking (see Fig.13.2).

Fig. 13.2   Office flythrough.

13.3.1.2   A virtual project office

The next stage in the development is to provide more information about the people within this world. One aspect is to be able to indicate whether people represented in the world are really available or not. A 'Rich Finger' server, or 'Activity Monitor' server is being developed to achieve this. The design draws inspiration from the 'Finger' service in Unix systems which allows one user to 'finger' other users to see if they are logged on, how active they have been recently and the state of their mailbox. The 'Rich Finger' will provide similar facilities for users on a range of different types of terminal. In addition to computer-based information, the system will also extract information from the private telephone system to indicate if they are on the phone now, or have been recently. The best way to present this information, to a user browsing a virtual office, is still a topic for research within the project, but could well include the use of a range of colours to indicate the degree of availability with pop-up windows giving fuller details, if required.

An important aspect of this approach to browsing is that the office does not need to represent a real office at all. The visualization could just as easily be used to represent a group of people who are working together but are actually distributed in many different places. One demonstration currently being developed is the Virtuosi project office. This will provide a 'Virtual open-plan office' with groups of desks representing the various partners in the project. Anyone working on the project will be able to enter the space and see who else in the project is available. The aim is to provide a much better feeling of working together than can be achieved normally with members of a project team located in many different places.

13.3.1.3   The fashion pilot

A different aspect of browsing for information is illustrated in the fashion pilot. In the fashion industry, as in many industries concerned with meeting consumers' demands, an important part of the design process is sourcing the various components of a product. In the clothing industry this may include yarns, textiles, accessories, etc. Depending on the position in the production chain, designers may need to assess the latest trends to incorporate in their designs, buyers may need to find suitable designers and manufacturers for their goods, and manufacturers may be looking for sub-contractors. Associated with the fashion pilot, Nottingham Trent University is developing a database of fashion information (see Fig. 13.3). The initial database is being constructed using a standard database package. However, studies are under way, to see how such an information brokering service could be extended to support the virtual organizations that are effectively set up every time a garment is designed, manufactured and sold. The development of such a service raises issues such as ownership of information, billing, efficient searching and retrieval, and many others.

As with other aspects of the project, 3-D may have a role to play. One possibility being investigated is representing the user interface to the information as a 3-D fashion trade show rather than a set of windows and menus (see Fig. 13.4). Not only would this make searching for information more acceptable for non-technical users, it would also enable users to see other people browsing the database, opening up the opportunity for them to discuss what they are seeing and to share discoveries.

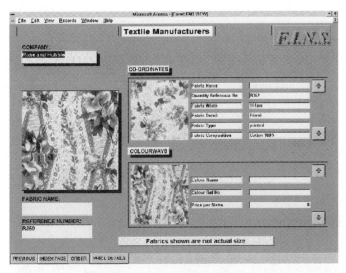

Fig. 13.3   A page from the fashion database.

Fig. 13.4   A virtual fashion show.

13.3.2   Viewing and presenting information

3-D and virtual reality also have a useful role to play in viewing and presenting information. The previous section has already shown that the 3-D paradigm is a very useful way of helping people to search for information. One reason for this is that humans are able to think and remember in spatial terms. This is also useful when presenting complex information because facts and trends that would have been hidden in a table of figures or even in a set of graphs can suddenly become very obvious when presented using a spatial metaphor.

The ideal way for collaborators to discuss a 3-D visualization of some information would be in a 3-D world. This has been investigated in Virtuosi. One example was a visualization of a database structure which became very clear when shown as a 3-D model (see Fig. 13.5). The model was generated in a virtual environment and two users were able to move about the structure and discuss its details.

Fig. 13.5   Database visualization.

One area very relevant to 3-D design is the fashion industry. The product of the industry is clearly a 3-D object. However, a garment is a very complex object. It is constructed from a number of 2-D shapes (the pattern pieces), but it does not have a shape purely on its own since it takes its shape from the person wearing it. Modelling a garment is thus a very complex problem involving the physical and mechanical properties of the cloth. Work is under way at Nottingham Trent University to model cloth and garments. The ultimate aim is to be able to create a virtual catwalk (see Fig. 13.6) with the mannequin and the garment moving in an entirely natural way. The computing power needed to achieve this will certainly be very great although not totally inconceivable, given current trends in computer development. More immediately, work is concentrating on achieving a realistic hang and drape on a stationary model. Tools are being developed to enable modifications to be made to a 3-D garment and then to recalculate the details of the garment. The 3-D garment will be available for viewing in a virtual world and it will be possible to share this view with a number of other people in the virtual world, all able to vary their own view point and to see the others who are sharing the world. To simplify the control of the mannequin, a voice recognition system is being developed so that users will be able to give spoken commands to direct the mannequin.

Fig. 13.6   The virtual catwalk.

This technology raises some interesting possibilities in the field of home shopping. Using these techniques it will be possible to construct the mannequin with the exact dimensions of an individual customer. The customer would be able to see remotely if the garment fits and, if necessary, the garment could be altered to fit exactly. The system could then produce bespoke garments by using the 3-D representation to produce the 2-D patterns that would be passed directly to the manufacturing unit to make the garment to the customer's exact specifications.

13.3.3   Interacting with people

In describing the use of 3-D for finding people and information and in presenting and viewing information, the fact that other users can inhabit the same virtual space has been mentioned a number of times. This opens up many new ways in which people may interact with each other remotely. One of the most important possibilities that affords itself is that people can use space to control their remote interactions in much the same way as they do in face-to-face meetings. Making a telephone call may be compared with knocking (loudly) on someone's door and entering when invited. However, many of our interactions are much more complex than this. On entering a room full of people we may choose who to speak to depending on who is already in conversation and with whom. We also find it useful to use the direction of a sound to help us monitor who is speaking and to listen to a particular conversation out of a number which may be going on.

To achieve this in a virtual world, a number of features must be provided. The first is to be able to represent other people in the virtual world so that they can be recognized (see Fig. 13.7). This does not necessarily mean that there needs to be a perfect likeness. Experiments within Virtuosi and other projects show that even a very basic block representation of a person can still give useful clues about other people in the environment and, with some visual indication when they are speaking, can enable useful conversation to take place.

Another obvious feature that is needed for interaction between people in the virtual world is the ability to hear each other. In the real world we do not hear the people in a large room equally loudly. Those close by are clearer than those further away and indeed some people will be seen but not heard. This has to be mimicked in the virtual world by controlling audio connections by the relative positions of users in the environment. Initial experiments have required audio connections to be set up as users enter a specific virtual world. However, an audio service is now being developed within Virtuosi which will automatically set up the connections when two or more people approach within a certain distance of each other. The architecture of the system is being designed to enable it to support either packet audio over the data network or the use of an overlay of telephone connections.

Fig. 13.7   A populated office.

An important aspect of the system design is the decision whether to use a fully distributed approach or to use a central audio server. The distributed approach with multi-casting of audio packets is the most flexible approach. However, in environments where multicast is not possible, a distributed system with a large number of users in a single conversation would require a large number ($\sim n^2$) connections. In these cases, a centralized server would be more practical.

Techniques are available which would enable not only the volume of the sound to be controlled but also its position in the virtual space to be fixed. This would add considerably to the complexity of the audio server, but would greatly enhance the feeling of really 'being there'. Virtuosi will be investigating spatialized sound and the benefits it can bring to enhance user interaction.

13.4   CONCLUSIONS

The Virtuosi project is tackling a range of issues concerned with providing support to groups of people working together. The use of virtual reality opens up a wide range of new ways of enabling people to interact with information and with other users. Using virtual reality it is possible to mimic the way in which we interact when we are in the same room much more closely than can be achieved with existing means of communication, such as the telephone or even videophone. A very important issue is integrating a range of different tools so that co-operation and sharing can be achieved as seamlessly as possible.

Although virtual reality can be used to mimic the real world, one of its major advantages is the increased control that users can have over their own environment. For instance, physical space is no longer a constraint and areas within the virtual environment can be expanded or contracted at will. It is also possible to create features such as 'secrecy zones' so that conversations that would otherwise be overheard, can be held in secret. An important area of research is to determine which facilities will be most useful in the virtual environment.

The Virtuosi project has been concentrating on presenting the virtual environment to users by means of conventional workstation displays. As the technology improves, there will also be opportunities to use fully immersive techniques for co-operative working. This will improve the realism of the interactions and increase the feeling of really 'being there'. The thought of putting on a VR headset and interacting with our colleagues, represented as angular blocks with eyes, seems rather strange at present, but the principles being investigated in Virtuosi may well lead to the new generation of computing and communications systems which will take the next step beyond multimedia.

REFERENCES

1.　Rogers A S: 'Virtuosi — support for groupworking', BT Technol J, 12, No 3, pp 81-89 (July 1994).

14

INTERACTIVE VISUALIZATION AND VIRTUAL ENVIRONMENTS ON THE INTERNET

G R Walker, J Morphett, M Fauth and P A Rea

14.1   INTRODUCTION

Three-dimensional, interactive graphics has the potential to liberate the human/ computer interface, and provide intuitive access to data landscapes and immersive applications — an appealing vision, which remains largely unfulfilled. The absence of widely accepted, cross-platform standards for distributed virtual environments and interactive visualization has restricted commercial applications to niche markets and specialist communities. Now, emerging industry standards, such as the virtual reality modelling language (VRML) and the Java programming language, promise to deliver the vision. These advances are supporting World Wide Web (WWW) applications with universal accessibility, and near-term developments will enable a wide range of Internet service offerings.

This chapter outlines the background to the emergence of VRML and Java in the context of current Internet development, and describes demonstrators which highlight the early capabilities of these standards — Portal is a VRML interface to a range of projects at BT Laboratories, while Jacaranda is an interactive Java visualization. Both demonstrators can be accessed on the BT Laboratories WWW site [1, 2]. The potential for future services is also considered, and it is concluded that support for interactive, multi-participant, multimedia environments and applications is an inevitable, short-term development of the current standards.

14.2   BACKGROUND

Bell *et al* have identified three stages in the historical development of the Internet
[3]. These are summarized in Fig. 14.1, starting with storage — the evolution of
the TCP/IP network infrastructure which provides a layer of abstraction between
data and physical machines. Stage two, retrieval, was the development of the
World Wide Web hypermedia system, built on the Universal Resource Locator
addressing scheme (URL) and the HyperText Markup Language document
standard (HTML). This made the distributed resources of the Internet more
widely accessible, and prompted rapid and sustained growth in network usage.
However, the overwhelming majority of material on the WWW remained passive
and two-dimensional, comprising, in effect, a vast multimedia database able to
send out pictures and text in response to requests.

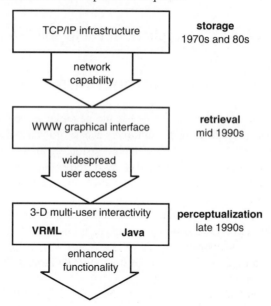

Fig. 14.1   Stages in Internet development (after Bell *et al* [3]).

This chapter is concerned with the third stage, perceptualization. An interface
to the Internet can be envisaged which is increasingly built around human interaction skills, and which delivers much more than remote data access. Perceptualization is the domain of this chapter's title — interactive visualization and virtual
environments. Many potential services require greater interface functionality,
including high-quality 3-D graphics and audio, complex user interaction — possibly tactile, and multi-participant capability. Developments such as VRML and
Java are starting to offer the capabilities required for this next stage, and they will
provide the building blocks for a range of compelling applications and services.

14.2.1   VRML

VRML was conceived in the spring of 1994 during a special interest group meeting at the first annual World Wide Web Conference in Geneva, Switzerland [4]. Several attendees described projects already underway to build three-dimensional graphical visualization tools which interoperate with the Web. It was agreed that these tools should have a common language for specifying 3-D scene descriptions and WWW hyperlinks — an analogue for virtual reality of the HTML standard. The term 'virtual reality markup language' (VRML) was coined, and the group resolved to begin work on a specification after the conference. The word 'markup' was later changed to 'modelling' to reflect the graphical nature of VRML. Within six months a draft standard was produced for review and a few months later the first version was released. It is intended that VRML should become the *de facto* language for interactive multi-participant simulation on the World Wide Web.

Released in June 1995, VRML 1.0 allows for the creation of virtual worlds with limited interactive behaviour. Figure 14.2 shows a segment from a VRML

```
#VRML V1.0 ascii
# a red cube translated 1 unit along the X axis with a link to a 'test-cube.html' file
Separator {
                    WWWAnchor{
                            name "http://iron.bt-sys.bt.co.uk:8080/jmorph/test-cube.html"
                            DEF redCubeSeparator{
                                Transform {
                                        translation            1 0 0
                                }
                                Material {
                                        diffuseColor           1 0 0
                                }
                                Cube {
                                        width 2
                                        height 2
                                        depth 2
                                }
                            }
                    }
}
```

Fig. 14.2   Extract from a VRML 1.0 scene description.

scene description. Although strongly influenced by Silicon Graphics, and in particular the Open Inventor file format, VRML was designed from the outset as an open standard, with key requirements of platform independence, extensibility and acceptable performance over low-bandwidth connections. VRML worlds can contain objects which have hyperlinks to other worlds, HTML documents or other valid multimedia file types. When the user selects an object with a hyperlink, the appropriate viewer is launched, e.g. Netscape. Similarly, when the user selects a link to a VRML document from within a correctly configured WWW browser, a VRML viewer is launched, e.g. WebSpace. VRML viewers are therefore a perfect complement to standard WWW browsers for navigating and visualizing the diversity of data on the Web.

Just two years on from the initial discussion in Geneva, there is now a plethora of application domains using VRML on the Internet. Table 14.1 (taken from the VRML Repository [5]) summarizes some of the categories for which VRML has provided a platform-independent method of transferring 3-D environments.

Table 14.1   Some of the categories for which VRML has provided a platform-independent method of transferring 3-D environments.

architecture	entertainment
art	environmental science
astronomy	history
biomedical sciences	home spaces
chemistry	maps and globes
commercial applications	mathematics
computer science	music
education	physics

Discussions on VRML 2.0 are well advanced, with significant contributors including Silicon Graphics, Sony, Apple and Microsoft, in addition to a range of universities and individuals [6]. An agreed revision to the standard is likely before the end of 1996, and proposals are focused on extensions to enable richer graphical worlds, interaction, animation and audio. Other enhancements could include explicit support for motion physics and real-time multi-user interaction.

Although VRML browsers are available for a range of hardware platforms, a relatively high-end PC or workstation is required for acceptable interactive performance. VRML models and applications are therefore not currently accessible to the majority of WWW users. Short-term developments in both hardware and software are certain to rectify this situation, and the standard will become increasingly widespread.

14.2.2   Java

Java is a programming language specifically designed for distributed-computing environments, and is a second key enabler for increased interest and interactivity of WWW applications [7-9]. Originally developed several years ago by Sun Microsystems as a control language for consumer electronics, Java was re-launched in early 1995. Figure 14.3 shows a segment of Java code. It is a derivative of today's standard object-oriented language C++, and includes a number of features which make it ideally suited to distributed Internet applications. These characteristics include:

- simplicity, especially in support of the Internet protocols required to access remote network resources — a corollary of simplicity is that Java programs, or applets, can be small and relatively easy to develop;

- secure, which is critical for a widely distributed and largely uncontrolled environment such as the Internet — once downloaded across a network, Java applets are dynamically interpreted within a secure environment with no access to local programs or data;

- architecture-neutral, enabling the same applets to run on any machine to which the Java interpreter has been ported — this is another essential attribute for a diverse environment such as the Internet, with a range of hardware platforms.

```
import java.applet.Applet
import java.awt.Graphics;

public class HelloWorldApplet extends Applet {

    public void paint(Graphics g) {
        g.drawString("Hello World!", 50, 25);
    }// end of method paint

}// end of class HelloWorldApplet
```

Fig. 14.3   Extract of Java code.

These and other features are not exclusive to Java, and competitors include VBScript [10] and ActiveX [11] from Microsoft, and General Magic's Telescript [12]. However Java was re-launched with perfect timing to meet market demand for a programming language for the Internet, and has already achieved widespread commercial acceptance. In the past year, Java has been acclaimed as the language of 'network-centric' computing. A range of Java applications, including spreadsheets and word processors can be summoned over the network as required, and companies such as Sun and Oracle are promising low-cost terminals or 'Internet appliances' optimized for the task. These developments raise important issues for the future shape of the entire computer industry, but this chapter is particularly concerned with applications to distributed visualization and virtual environments. In this context, it is probable that VRML 2.0 will support Java as a preferred scripting language, thereby complementing VRML graphics with the power and flexibility of Java programming.

Netscape, the most popular WWW browser already supports Java applets, highlighting the remarkable pace of development in Internet products and standards. Moreover, several comprehensive commercial Java development environments are available, with programmer support increasing rapidly. The interactivity of Java is therefore already familiar to many Internet users, and the range of applications is growing daily.

14.3   APPLICATION DEMONSTRATORS

The previous section introduced the key features and current status of VRML and Java, suggesting that they provide the functionality required for new interfaces and services which will result in perceptualization over the Internet. This section describes demonstrators which highlight early capabilities of the standards and point to future service opportunities.

14.3.1   Portal

Portal provides an introduction to some of the work and facilities at BT Laboratories (BTL), and is an example of an Internet virtual environment. It is accessible from the BTL WWW server [1], and is both an evolving and an involving interface to BT's projects and other visitors (Fig. 14.4). As described in this chapter, it has been implemented using the VRML 1.0 standard.

Fig. 14.4   Portal interface.

Portal is prefaced by a three-dimensional model of BT Laboratories (see Fig. 14.10 in section 14.4), which illustrates the ability to present physical environments using VRML and which points to future applications in planning and design. The user is then invited to select a personality icon, which provides the user with a choice of persona within Portal:

- business;

- human factors;

- marketing;

- education;

- technology.

Once a persona is adopted, all subsequent information can be tailored to the individual. For example, a user who selects the human factors personality icon might be informed of the psychological, sociological and ergonomic aspects of the projects they visited.

The projects are displayed as islands (Fig. 14.5) and are currently:

- London Model — an area of central London, with potential radio planning applications;

- Fly the Network — a three-dimensional network management interface;

- Electronic Agora — a user-centred videoconferencing system;

- CamNet — a mobile, ISDN telepresence system;

- VISA — an interface to home services;

- Workspace 2000 — a physical desktop of the future.

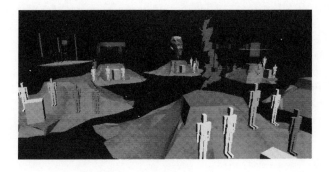

Fig. 14.5   Portal project islands.

From knowledge of their personality icon, individuals are represented by automatically placing a colour-coded embodiment of them on the project islands they visit — Portal people. This aids in focusing information for subsequent visitors. They will see a presence on the islands, which indicates both the popularity of an island and the personality of other visitors, from the number and colour of the embodiments, respectively. In addition to colour and placement, transparency is used as a means of temporal presence, so that embodiments 'fade away' over time. This use of embodiments provides a sense of sharing and association within the space of Portal, in contrast to the normal sterile isolation of WWW information. For the Portal visitor, there is a sense of association which extends beyond simple multimedia data retrieval.

This feature is achieved through combining Perl scripts with the VRML language, an extremely powerful combination of Internet resources. Perl is a language designed for manipulation of text, files and processes, and it has been used extensively to manage and build intelligence into WWW sites. It has been suggested that Perl could act as the behaviour language for VRML [13]; however, using a script in this manner does not provide truly interactive behaviour, since the VRML environment remains static once loaded into the browser. While this latency is acceptable in applications such as Portal people, future applications will inevitably demand full interactivity and synchronous communication. This could be provided, as has already been noted, through the ability of Java to manipulate the content of VRML scenes.

Within the Portal islands the user can fly to pre-set viewpoints, using a 'Viewpoints' menu option within the browser. This provides a macro level of navigation, enabling the world developer to assign cameras to strategically important or interesting views of the world. Once a viewpoint is selected, the user is flown to that position and is then free to navigate the world at a microlevel, using standard VRML browser controls. By maintaining a sense of direction and travel, flying to information reduces the problem of getting lost in jumps between hyperlinked data. Current browsers also support limited interaction within the VRML scene — when the cursor passes over a hyperlink, the object is highlighted and if the mouse button is clicked, the associated URL is retrieved.

In the current version of Portal, the links from the VRML worlds are to HTML pages relevant to the selected object. The pages give a textual outline of the projects, and are in turn hyperlinked to further three-dimensional VRML models (Figs. 14.6 and 14.7). It has also been proposed that such links could provide the basis of asynchronous communications (such as Email) within the Portal virtual environment [14].

Fig. 14.6   London model — a 3-D environment for line-of-sight radio planning.

In the first six months following the release of Portal in July 1995 almost one hundred thousand access requests were registered. Allowing for the limited availability of VRML browsers, such statistics emphasize the power of the Internet in reaching a wide audience. The wide diversity of the user community is also confirmed, with requests ranging from Brighton to Bolivia.

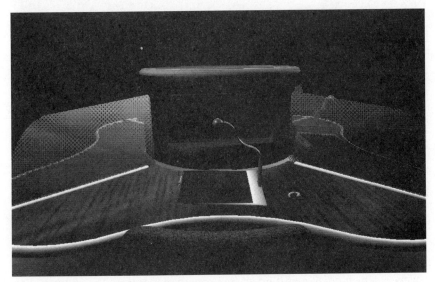

Fig. 14.7   Workspace 2000 — flexible support for remote collaborative work.

14.3.2   Jacaranda

Jacaranda (Java call reporting and analysis) illustrates the potential for Java to provide interactivity within Internet-based visualization applications [2]. The Jacaranda demonstrator conveys a vision of future developments in on-line service reporting and management, in which Java is used to deliver the results of a database query together with an interactive visualization application. The end-user is able to explore the underlying data, and to interact with a visual display of telephone-call record statistics.

Jacaranda is an interactive visualization of telephone calls made from within the UK to a single enquiry point, for example a product-support line or customer enquiry number. The geographical source of calls is represented by towers located at the main centres of population. The height and colour of a tower is mapped to the number of calls originating from that town in a single hour (Fig. 14.8).

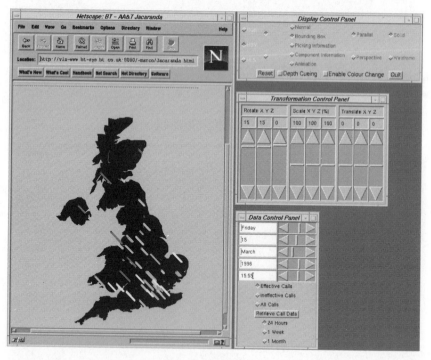

Fig. 14.8   Jacaranda interface — overview.

The user can select the date and time of interest, and initiate animation sequences which might reveal interesting patterns and trends. The ability to link from selected towers to supplementary information is demonstrated, and it is also possible to rotate and scale the 3-D visual display. The application comprises four primary windows (Fig. 14.8) — the main visualization, the display control panel, the data control panel, and the transformation control panel.

The display control panel is used to select the overall mode, while the transformation control panel allows the user to scale, rotate and translate the visual output. The data control panel selects the call records to be displayed and controls the animation function. A fifth window, the 'pick information' panel provides more detailed data at a selected location (Fig. 14.9). An extended description of the application can be found on the WWW pages [2].

Fig. 14.9   Jacaranda interface — 'pick information' panel.

The functionality of Jacaranda is almost identical to a visualization concept demonstrator which was implemented over two years ago using a commercial software development environment for stand-alone visualization applications [15]. However, the ability to deliver this functionality over a network to any terminal with a Java-enabled browser offers important practical advantages, including:

- wide base of installed and configured WWW clients — the application developer need not be concerned with either the hardware or software of the user environment, and hence incurs greatly reduced client set-up and support costs;

- platform independence eliminates problems and costs of porting — there is no need to write and maintain separate versions for Macintosh, Unix, NT, Windows 95, etc;

- single current version on the central server — reduced costs of version control and distribution, and immediate, universal upgrade and bug-fix capability.

Jacaranda was first developed in the Spring of 1995 using the alpha release of Java at a time when there was minimal developer support. It was therefore necessary to code all the 3-D graphics and interface classes, and the final application was only compatible with a limited range of WWW browsers. Jacaranda has recently been updated to Java 1.0, making it more widely accessible. The diagrams in this chapter are taken from this latest implementation.

14.4   FUTURE SERVICES

Our initial concept demonstrators, Portal and Jacaranda are only scraping the surface of the service functionality that will shortly be available as standards and support tools continue to develop. This section outlines a number of potential applications in interactive visualization and virtual environments, and includes brief consideration of Internet audio developments. The applications are grouped into categories requiring increasing functionality.

14.4.1   Three-dimensional models

The current VRML standard permits the construction of three-dimensional models, and provides pre-set viewpoints in addition to user fly-through. It is suggested that standard libraries might be used to distribute textures and other bandwidth-hungry details which help to provide photo-realistic scenes. Specific services might include:

- on-line product catalogues or virtual shopping mall;

- architectural walk-throughs (Fig. 14.10), and public review and comment on proposed urban developments;

- inclusion of three-dimensional 'figures' within electronic publications and on-line manuals.

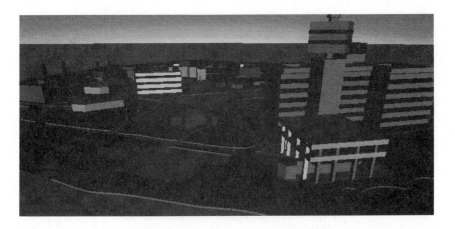

Fig. 14.10   BT Laboratories site model.

14.4.2   Modelling and user interaction

The inclusion of support for interpreted scripts is a certain near-term development in the VRML standard. This will enable the functionality and interactivity of Java applets (and potentially other languages) to be included within VRML worlds. No longer will the downloaded worlds remain static and sterile; objects will be brought to life with preprogrammed behaviours and responses, and independent applications will have their own existence within the world. The user will be able to interact with menus and control panels, just as they would in a conventional computer interface. Potential services include:

- educational experiments and simulations;

- interactive data visualizations and modelling (Fig 14.11);

- remote control and navigation interfaces;

- interactive adventure games.

Fig. 14.11   Visualization of neural network channel assignment algorithm [15].

14.4.3   Multi-user environments

A further extension to server and browser functionality would enable multi-user environments or shared spaces [16, 17]. Such applications are particularly attractive to service and network providers, in that participants contribute to the interest and 'content' of the world, thereby reducing the demand for regular updates, revisions, and costly new material. Parallels can be drawn with telephone conversations, in which the service provider supplies an 'empty' audio connection, and all content is generated by the participants.

A multi-user, interactive space could support services such as:

- multi-participant games;

- participatory review of product or building designs;

- a virtual office and communications interface for a distributed project team;

- a general-purpose environment for CSCW (computer-supported co-operative working) applications (Figs. 14.12 and 14.13).

In addition to increased security concerns, multi-user applications must also address issues of scalability, with respect to both server performance and network traffic [18-20]. Further discussion is again beyond the scope of this chapter, but there is a range of technical challenges in developing a system architecture to support the required functionality, particularly in areas such as service creation and billing which have received minimal attention in early demonstrators.

Fig. 14.12   Physical CSCW environment.

14.4.4   Audio

So far this chapter has concentrated on extending the functionality of the visual interface. However, the critical contribution of audio must also be considered, and in particular the role of spatial audio in enhancing user experience of virtual environments. Recent Internet developments include RealAudio, which streams audio with real-time decompression [21], and Internet Phone which provides a full duplex audio link, albeit with unacceptable delay for most applications [22]. Future developments will support spatial audio within distributed applications. This will include sound as an aspect of object behaviour, discrete audio events, and also user input in the form of speech. Such functionality will greatly enhance applications such as games or shopping, and be an essential component of CSCW environments.

Fig. 14.13   Virtual CSCW environment [19].

14.5   DISCUSSION

This chapter has shown that current developments in Internet standards will promote a transformation in applications. Jacaranda and Portal provide only an early indication of the diversity of services that will be supported by future server and browser functionality. Moreover, with standards will come the critical mass required to grow major new markets out of current trials and niche applications. These developments will lead to currently unimagined services, with far-reaching implications for the Internet, at least equal to the changes resulting from the development of the WWW. Potential for a profusion of new services notwithstanding, such developments also raise important commercial and network-related issues. In this section some of the key considerations are briefly discussed, with particular attention to the role of the network provider.

While a degree of certainty can be claimed regarding near-term technical developments, commercial Internet developments are largely unpredictable even on a time-scale of months. Anticipating the successful services or 'killer applications' entails a significant element of speculation, and although there is some scope for major companies to influence developments, the essentially open and uncontrolled nature of the Internet will ensure a continuing and healthy diet of the unexpected. The market is already a complex tangle of alliances, joint ventures, established computer and telecommunications companies, and promising start-ups.

Although an unpredictable wealth of services will come and go in this unstable market, network providers will nevertheless need to provide guarantees on quality of service, facilities for security and billing, and perhaps a common infrastructure of network-based storage and processing capability. In parallel with the emerging content and application standards such as VRML and Java, the development is being seen of new network protocols which are designed to meet the traffic requirements of selected services. Examples include RSVP (reservation set-up protocol) which allows Internet applications to obtain a specified quality of service, typically by reserving resources along the data path [23], and RTP which provides support for applications with real-time properties, including timing reconstruction and loss detection [24].

However, the distributed approach to management and ownership of the Internet will inevitably give rise to critical issues which are essentially 'everybody's problem', and yet are simultaneously 'nobody's problem' when it comes to prescribing a solution. Examples include service-impairing congestion, and blocking or filtering of pornographic material. Such considerations will create a role for Internet service providers to maintain islands of high-quality access, which would include an editorial function to filter and catalogue the wealth of available information. They will also provide scope for the same cost-effective IP network technology to be used within the managed environment of a closed user group on a private network — so-called Intranets.

One secure prediction for the future Internet services is that market instability and commercial uncertainty will endure, and the pace of change will be sustained or even accelerate. Committing to traditional type is a hazardous undertaking, as evidenced by the overwhelming majority of references in this chapter, which are only available in electronic form. This simple observation is an early example of the shifts in business and working practices which have been provoked by early Internet applications, but once again is only a taste of the changes yet to come.

14.6   CONCLUSIONS

A key reason for the success of the Internet is that the protocols and standards are as generic and low-level as is practicable. These sturdy building blocks provide powerful enablers to the wider community of application developers.

Returning to the stages in Internet development outlined in Fig. 14.1, it was shown that stage two, retrieval, was built on URL addressing and the HTML document standard. This is a very basic level of functionality, and yet has spawned an abundance of applications and universal interest in Internet services.

The increase in functionality offered by developments such as VRML and Java will lead on to the third stage, perceptualization, transforming the interface and enabling a wealth of new services, far in advance of current WWW browsers and existing passive Internet databases. Early demonstrators such as Portal and Jacaranda provide a glimpse of future services, which will include interactive visualization and multi-participant virtual environments.

APPENDIX

Glossary

As with any technical discipline, the Internet has developed an extensive vocabulary of acronyms, familiar and convenient to those working in the area, but superficially intimidating to newcomers. It would have been impractical to avoid these terms in this chapter, but the following glossary (based on http://www.cwru.edu/help/webglossary.html) is offered in mitigation.

Applet — a small Java application embedded within a WWW page, and requiring another Java program (such as a browser) in order to run.

Browser — software that allows you to navigate information databases; examples are Netscape Navigator and NCSA Mosaic.

HTML — HyperText Markup Language, used to tag various parts of a WWW document so browser software will know how to display that document's links, text, graphics and attached media.

IP — Internet Protocol, a set of standards that control communications on the Internet. An IP address is the number assigned to any Internet-connected computer.

Java — a programming language specifically designed by Sun microsystems for distributed computing environments.

Perl — a programming language designed for manipulation of text, files and processes.

Protocol — a set of standards that define how traffic and communications are handled by a computer or network router.

TCP/IP — 'transmission control protocol/Internet protocol' — the basic protocol controlling applications on the Internet.

URL — Uniform Resource Locator, the addressing system used in the WWW and other Internet resources. The URL contains information about the method of access, the server to be accessed and the path of any file to be accessed.

VRML — Virtual Reality Modelling Language, intended to support interactive multi-participant simulation on the World Wide Web.

WWW — World Wide Web, a hypertext-based Internet service used for browsing Internet resources.

REFERENCES

1. BT Labs Portal, http://virtualbusiness.labs.bt.com/vrml/portal/home

2. BT Labs Jacaranda, http://www.labs.bt.com/innovate/informat/jacaranda

3. Bell G, Parisi A, and Pesce M: 'The Virtual Reality Modelling Language: version 1.0 specification', http://www.wired.com/vrml.tech/vrml10-3.html (1995).

4. Raggett D: 'Extending WWW to support platform independent virtual reality', http ://vrml.wired.com/concepts/raggett.html

5. VRML repository, http://www.sdsc.edu/vrml

6. Honda Y, Matsuda K, Bell G and Marrin C: 'An Overview of the Moving Worlds Proposal for VRML 2.0', http://webspace.sgi.com/moving-worlds/over3/overview main.html (1995).

7.   Gosling J and McGilton H: 'The JAVA language environment: a white paper', http://java.sun.com/whitePaper/javawhitepaper_1.html (1995).

8.   Sun microsystems Java WWW pages, http://java.sun.com/about.html

9.   JavaWorld, on-line Java magazine, http://www.javaworld.com

10. Microsoft VBScript (Visual Basic scripting language), http://www.microsoft.com/vbscript/vbsmain.htm

11. Microsoft ActiveX (Activate the Internet development kit), http://www.microsoft.com/corpinfo/press/1996/mar96/activxpr.htm

12. General Magic Telescript, http://www.genmagic.com/Telescript/

13. Moreland J L and Nadeau D R: 'The Virtual Reality Behaviour System (VRBS): Using Perl as a behaviour language for VRML', Proc 1995 VRML Symposium, San Diego (November 1995).

14. Morphett J: 'Presence in Absence: Asynchronous communication in virtual environments', Proc 3D and Multimedia on the Internet, WWW and Networks, Bradford, UK (April 1996).

15. Walker G R: 'Challenges in Information Visualization' Br Telecomunications Eng J, 14, p 17 (April 1995) and http://www.labs.bt.com/innovate/informat/infovis

16. Bradley L, Walker G R and McGrath A: 'Shared Spaces', Br Telecomunications Eng J, 15, (July 1996) and http://virtualbusiness.labs.bt.com/SharedSpaces

17. Honda Y, Matsuda K, Rekimoto J and Lea R: 'Virtual Society: extending the WWW to support a multi-user interactive shared 3-D environment', Proc 1995 VRML Symposium, San Diego (November 1995).

18. Sony Virtual Society project, http://vs.sony.co.jp/VS-E/vstop.html

19. Greenhalgh C: 'MASSIVE teleconferencing system, http://www.crg.cs.nott.ac.uk/~cmg/massive.html

20. Swedish Institue of Computer Science, Distributed Interactive Virtual Environment project, http://www.sics.se/dive

21. RealAudio, http://www.realaudio.com/

22. Internet Phone, http://www.vocaltec.com/iphone.htm

23. RSVP overview, http://www.isi.edu/div7/rsvp/rsvp.html

24. RTP overview, http://www.fokus.gmd.de/step/employees/hgs/rtp/faq1.html

15

YACHT VIDEO SYSTEM FOR THE WHITBREAD ROUND THE WORLD RACE

C D Woolf and D A Tilson

15.1   INTRODUCTION

The Whitbread Round the World Race is an international yachting event held every four years. This tough and dangerous event attracts major sponsors who demand media coverage throughout its nine-month duration. When the yachts are in the vicinity of the shore, conventional techniques can be used to obtain up-to-date broadcast video and audio material. These include 'chase boats', line-of-sight microwave links (sometimes via an overhead helicopter to extend coverage) and the dropping into the water of watertight vessels containing video tape for later retrieval.

When the yachts reach mid-ocean the broadcasters previously relied on old library footage to support the news from the voice and text messages, received via HF radio and satellite. The lack of live, or recently received, supporting video material tended to reduce the impact of the stories unfolding at sea, leading to a general decline in interest in the race until the yachts neared the next port.

The Yacht Video System (YVS) was developed to enable the yachts to relay near-live video and audio material from mid-ocean at a quality acceptable for broadcasting. The system uses video compression, satellite communications, and data 'store-and-forward' techniques with a bespoke error-correction protocol.

This chapter initially provides background on BT's involvement in the 1993 race, followed by discussion of the results collected during system trials. A detailed system description is given, with particular attention to error correction. This is concluded with a brief statement of performance and results.

15.2   RACE INVOLVEMENT

The 1993 Whitbread Round the World Race was the second time that BT had run the communications for the race, providing media centres at each port of call (Southampton, UK; Punta del Este, Uruguay; Fremantle, Australia; Fort Lauderdale, USA). As well as the usual telephone and facsimile facilities, BT developed two technology platforms — the Race Results System (RRS) providing the media with six-hourly yacht-position reports presented graphically on a PC as route maps, and the Yacht Video System (YVS) described in this chapter.

Sponsorship of the race enabled BT to:

● demonstrate a major showcase of existing products and services on a global basis;

● provide global visibility of its corporate identity;

● show that it is at the forefront of leading-edge technologies and how these can be developed into products and services for the future.

All fifteen of the yachts were equipped with RRS, whereas only ten yachts had YVS due to cost and support limitations.

15.3   SATELLITE COMMUNICATIONS

Several forms of radio communications are used to communicate with yachts racing in mid-ocean. However, only mobile satellite communication is capable of providing reliable 24-hour coverage over a wide area. It is certainly the most flexible and cost-effective method of providing the bandwidth necessary to transmit video regularly from yachts in mid-ocean.

A review of available satellite systems clearly demonstrated that only Inmarsat could provide the communications capacity required. Inmarsat is an international treaty organization with BT as its UK signatory. It was established in 1979 to provide satellite communications to the maritime community and has evolved to become the only provider of global mobile satellite communications for commercial, distress and safety applications at sea, in the air and on land.

As shown in Fig. 15.1, Inmarsat's coverage is global (except for the polar regions) using four satellites to provide communications for the maritime, aeronautical and land-mobile communities. The Inmarsat-A service was chosen for the YVS due to its ability to transmit large quantities of data efficiently.

Inmarsat-A is used extensively for routine communications by the world's maritime community and provides voice, telex, facsimile and voice-band data

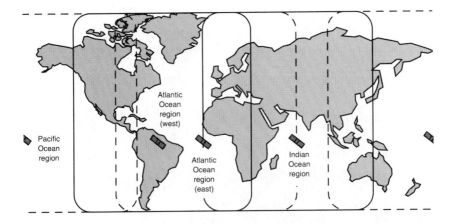

Fig. 15.1 Inmarsat coverage.

facilities. In addition, a high-speed-data (HSD) option supporting both simplex and duplex communications at 56/64 kbit/s data has become available in recent years.

All Inmarsat services operate in the mobile satellite service (MSS) bands, around 1.5 GHz (space-to-earth) and 1.6 GHz (earth-to-space). Inmarsat-A ship earth stations (SES) use an antenna of about 1 m diameter to achieve the required signal-to-noise ratio in both transmit and receive directions.

Frequency modulation (FM) is used for the voice, facsimile and voice-band data services. The HSD option utilizes a digital modulation scheme known as quadrature phase shift keying (QPSK) along with powerful forward-error-correction techniques to enhance the error performance of the link.

The simplex HSD version provides a high-speed channel from the yacht to the shore along with a return 'audio only' channel back to the yacht. This audio channel was used in the YVS to send the small amounts of data necessary for the operation of the error-correction scheme described in section 15.7.

The nature of installations on ocean racing yachts required the identification of an SES that was small and lightweight, with modest power consumption. The SES chosen (MTI MCS-9120) included a computer-controlled mechanism to keep the 0.9 m diameter antenna pointed towards the satellite. Yacht designers were able to incorporate the antenna and supporting equipment in either the fore or aft watertight compartments. Figure 15.2 shows the installation of the equipment on-board the Maxi class yacht Merit Cup, and is typical of installations on the other yachts taking part in the race.

Fig. 15.2   Maxi yacht with deck-mounted antenna radome.

15.4   STORE-AND-FORWARD PRINCIPLE

Providing live broadcast material from the yachts would require the video and audio to be compressed from the theoretical maximum data rate of 140 Mbit/s to the 56/64-kbit/s limit of the satellite channel (approximately 2000:1 reduction). Although equipment is readily available to achieve these high levels of compression (typically 48 kbit/s for the video and 16 kbit/s for the audio), the resulting image quality is unsuitable for television broadcasting. Using store-and-forward techniques the level of compression is not directly limited by the channel capacity but by factors such as transmission time. The system then becomes more analogous to conventional file transfer over a network, where the data is video and audio. This means that the material can be stored at a data rate higher than the satellite channel capacity, enabling the video quality to be improved dramatically. Although the resulting material is not strictly live, it could be only a matter of minutes old.

15.5   SYSTEM OVERVIEW

The video material is collected from a combination of fixed camera positions (mast, stern, navigation cockpit, etc) and crew members using camcorders. This is edited down to a 'video clip' of a few minutes duration with bespoke editing suites based on varying techniques such as back-to-back camcorders, handheld video players, and full-size video cassette recorders. Once the clip is prepared it is played through the YVS which compresses the material in real time using the H.320-compliant [1] BT VC2300 video and audio codec, developed at BT Laboratories. The resulting data is stored internally ready for transmission.

A simplex high-speed-data (HSD) call is then initiated on the Inmarsat-A SES to the on-shore receiving equipment (designed to operate unmanned). The crew member requests an HSD call through the appropriate land earth station and satellite (e.g. Goonhilly at 64 kbit/s through Atlantic East). The call is then routed via the UK ISDN2 network to the automatic receiving equipment. The primary collection point for the clips was Reuters (London), although a back-up service was provided at BT Laboratories (see Fig. 15.3).

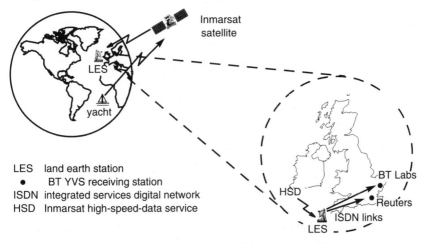

Fig. 15.3 Communications overview.

Once the yacht-to-shore data link is established, the YVS transmits an information header, including a yacht identifier and clip details, and waits for acknowledgement from the receiving system. A successful handshake is followed by the yacht transmitting the complete video clip for storage on shore.

Establishing the yacht-to-shore data link provides a return 'audio-only' channel for error-correction purposes. The audio channel is used by the shore system to provide a simplex modem-based data link for requesting the yacht system to retransmit data corrupted during transmission. The transfer is complete when all the data has been received without error and the yacht system informed. The shore system stores the clip for later recovery and prepares itself for another call.

During the day the receiving system is accessed to retrieve the new clips, either routinely or by specific request from the broadcasters. A directory of clips is displayed enabling selection by the operator. Once selected, the clip is then played back at the original data rate through the BT VC2300 codec, and the resulting video and audio recorded for subsequent editing and onward dissemination.

15.6   SYSTEM TRIALS

Using a prototype YVS system, extensive tests and trials were performed at BT Laboratories, and on Maxi (Merit Cup) and W60 (Yamaha) class yachts. This was to determine optimum system configuration for the environment in which the final design was to operate. The trials included broadcasters, producers, engineers and yachtsmen to ensure the opinions of all interested parties were considered.

15.6.1   Data rate selection

Selecting an appropriate compressed video and audio data rate depended on:

- quantity of video and audio material to be stored;

- resulting quality after compression;

- time taken to transmit successfully;

- satellite channel integrity (including bit errors and outages).

The consensus was to use a data rate of 768 kbit/s for the compressed video and audio, with a maximum storage time of 2.5 min of original material. This quantity of material compressed at 768 kbit/s would take approximately 30 min to transmit over a 64-kbit/s satellite channel (excluding error correction), which is an increase by a factor of twelve in transmission time against material time. In the event of bad transmission conditions a 384-kbit/s option was considered necessary as it halved the transmission time (an increase by a factor of six in transmission time against material time), but with slightly lower video quality.

Selecting G.722 [2] wideband audio (7 kHz bandwidth at 48 kbit/s) ensured the video clips contained good-quality sound, as in many situations this was considered of prime importance.

15.6.2   Channel performance

The Inmarsat-A HSD satellite communications channel operates with an error rate better than one error in 10^6. This level of performance is within the error-correcting capabilities of the video codecs and would normally be satisfactory. However, tests on board the yachts highlighted problems created by the highly dynamic motion encountered at sea affecting the SES.

As the yacht moves the antenna tracks the satellite to ensure that a good connection is maintained. A number of sensors are used to determine the motion of the yacht:

- vertical reference sensors (pitch and roll);

- flux gate compass (yaw);

- accelerometer (rate of change).

The data from these sensors is used to adjust antenna azimuth and elevation to maximize the received signal strength. During sea trials it was found that it was quite possible for the motion of the yacht to be so violent that the antenna was not able to remain locked on to the satellite. The resulting outages could be up to 40 sec in duration, causing significant sections of compressed video and audio data to be lost. Even a short outage would result in a momentary loss of video and audio, followed by significant degradation of the video material for several seconds when played back on shore. This was due to the high levels of compression applied by the codec.

It was clear from the tests that a unique error-correcting system would be required to handle the data outages, and to accommodate the characteristics of the satellite channel (long delays and an 'audio-only' return channel).

15.6.3   Environment

The carbon construction of the Maxi-class yacht required the radome of the satellite dish assembly to be situated above decks to enable transmission. Located in the fore or aft deck as a plastic 'blister', the radome was supported about its circumference with reinforcing to absorb the deck flexing. The W60 class yacht used kevlar which did not have a significant effect on signal strength and enabled the dish assembly to be mounted inside the hull, preserving the integrity of the deck and protecting it from the elements.

Data from previous races and the trials clearly demonstrated that the yacht system would have to survive a very hostile environment comprising:

- sea-water ingress from storage of wet sails and crew member's clothing;

- condensation (especially on the bulkheads);

- vibration (principally caused by the keel oscillating as it cleaves the water);

- shock caused by yachts falling off waves or hitting them head on (up to 2g);

- temperatures ranging from around − 5 °C to 45 °C;

- high humidity (approaching 100% in the tropics).

In addition, reliability had to be very high as system failures could not be easily rectified.

15.6.4   Safety

Safety was a very important issue as the SES is basically a microwave transmitter situated in close proximity to the crew. A number of measures were taken to prevent exposure during transmission:

- microwave absorbing foam on bulkheads leading to crew quarters;

- markings on deck to indicate safe working distance;

- purpose-built safety system comprising warning indicators and stop buttons situated at key positions on the yacht (with manual override for emergency situations).

15.7   ERROR-CORRECTION PROTOCOL

The initial trials of the YVS identified the devastating effect of channel outages on the quality of video material. Some form of automatic repeat request (ARQ) protocol was required to ensure that all the data was transferred correctly despite the outages caused by the motion of the yacht.

Although many forms of ARQ protocol have been developed none were found that satisfied all the requirements of this particular application. The main deficiency was the inability of standard protocols to maintain a connection during long outages while still making efficient use of the communications capacity when it was available. To overcome these deficiencies a new protocol was designed to meet the special requirements of the YVS.

15.7.1   Types of ARQ protocol

There are several types of ARQ protocol that have different levels of performance and implementation complexity.

- Stop-and-wait represents the simplest ARQ procedure. The transmitter sends a packet of data and waits for an acknowledgement (ACK) from the receiver. If the ACK is not received within a specified period (time-out) the packet is retransmitted. When the ACK is received the next packet is transmitted. Stop-and-wait protocols, while easy to implement, can be very inefficient particularly for satellite channels.

The other main family of ARQ protocols send data continuously. As there is no idle time spent waiting for individual ACKs, these protocols are inherently more efficient. There are two main types.

- Go-back-N retransmits data from the point where an error was detected. Go-back-N is much more efficient than Stop-and-wait, but may retransmit data that had been received correctly. This ARQ system can result in throughput degradation if a large round-trip delay is involved, as with satellite channels.

- Selective-repeat also sends data continuously. However, it can be more efficient than a Go-back-N protocol as it only requests retransmission of individual packets received in error. With this scheme, a buffer must be provided at the receiving end to store the error-free packets arriving after a packet with errors. This is necessary to preserve sequence integrity. To restrict the amount of buffering required, the number of packets that can be stored is usually limited. This limit is referred to as the window size. To ensure that the channel is used efficiently in error-free conditions the window size must be at least the same size as the amount of data that can be transmitted in a round-trip delay (the bandwidth/delay product).

The ARQ protocol developed for the YVS is based on the selective-repeat principle. It is unusual in that it uses an effectively infinite window size and has time-out values specifically chosen to cater for the peculiarities of the mobile satellite channel used.

15.7.2   YVS protocol structure

By their nature ARQ protocols require a duplex channel. In the YVS system the yacht-to-shore channel was provided by the 56/64-kbit/s HSD link. The channel in the return direction was supported by an audio channel carrying voice band data (VBD).

The format of the data structures in each direction was determined by the characteristics of these channels and the functions required. The YVS protocol was defined in terms of these data structures and are described in Table 15.1.

15.7.2.1   Yacht-to-shore data blocks

The structure of the data blocks transmitted from the yacht is shown in Fig. 15.4. The 'preamble' is 8 bytes of alternating binary ones and zeros which indicate the imminent arrival of a data block. The 'preamble' together with the 'unique word' (of known contents) allows the receiver to perform byte synchronization, and therefore determine the position of the other fields.

Table 15.1   YVS protocol data structures (a) yacht-to-shore (b) shore-to-yacht.

(a) Yacht-to-shore	Description
Data block	Contains the audio and video material
Transmission header block (THB)	Used at the start of a transmission to identify the yacht, the video coding rate and the size of the video clip
(b) Shore-to-yacht	Description
Ack THB	Used to acknowledge receipt of the THB
Rq Blks	Used to request retransmission of specific data blocks
Ack All	Used to acknowledge the successful reception of all the data blocks

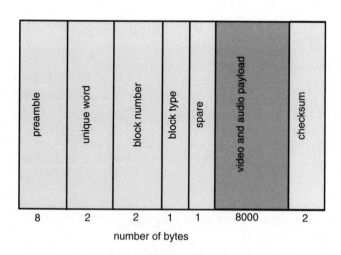

number of bytes

Fig. 15.4   Yacht-to-shore data blocks.

The 'block number' provides a unique numerical identifier for each block. The 'block type' field distinguishes between data blocks (containing video) and the special transmission header block (THB) containing yacht identifier, coding rate and clip length. A 'spare' 1-byte field was included to cater for any enhancements that were found to be necessary during implementation.

The default size for the data field is 8000 bytes. This value was chosen to ensure that transmission efficiency was maintained. The transmission time (about 1 sec) is also short enough to limit the amount of data to be retransmitted in the event of an error. A shorter data field of 400 bytes was used in the THB.

The 16-bit 'checksum' is used to detect when a block is received in error.

15.7.2.2   Shore-to-yacht packets

The structure of the packets transmitted from the shore is shown in Fig. 15.5. The 'unique word' (of known contents) is used to indicate the start of a packet. A preamble is not required in the shore-to-yacht direction as the asynchronous VBD channel supports byte-oriented data.

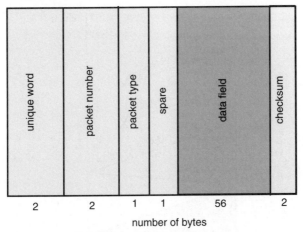

unique word	packet number	packet type	spare	data field	checksum
2	2	1	1	56	2

number of bytes

Fig. 15.5   Shore-to-yacht packets.

The 'packet number' field provides a unique numerical identifier for each packet. 'Packet type' distinguishes between the three types of packet that can be present on the shore-to-yacht channel (Table 15.1(b)). A 1-byte 'spare' field was included to cater for any enhancements that were found to be necessary during implementation.

The 56-byte 'data field' is only present in the packets used to request the retransmission of data blocks. It is used to request the retransmission of up to 28 blocks. The 16-bit 'checksum' is used to detect when a packet is received in error.

15.7.3   YVS protocol — operation

Once the video and audio material has been stored in YVS it is ready for transmission ashore. The YVS ARQ protocol has three distinct phases.

15.7.3.1   Initialization phase

After communication has been established the yacht sends the transmission header block (THB). The THB contains information needed by the shore-based receiving station. The shore station acknowledges successful reception of the THB by sending an Ack THB packet. The yacht starts to send data blocks (containing the audio and video data) when this acknowledgement is received correctly.

If the Ack THB does not arrive when expected, the THB is retransmitted periodically. Similarly the Ack THB is retransmitted if the first data block does not arrive when expected. This is illustrated in the time sequence diagram in Fig. 15.6.

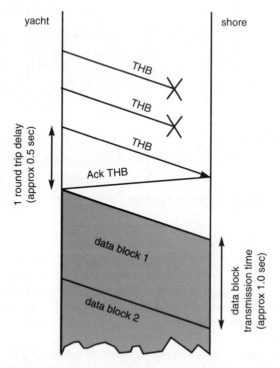

Fig. 15.6   YVS-error correction protocol — initialization phase.

15.7.3.2   Data transfer phase

The shore receiving station performs error checking on each block as it is received, before storing it in memory. Blocks identified as corrupted are noted in a table for later correction. No communication with the yacht takes place during this phase. When all the data has been transferred the error-correction phase commences.

15.7.3.3   Error-correction phase

The shore receiving station uses a Rq Blk packet to request the retransmission of up to 28 data blocks that were received with errors or were not received at all. On receipt of a Rq Blk packet the yacht starts to retransmit the requested blocks. The same Rq Blk packet is retransmitted periodically until the first of the retransmitted data blocks arrive.

The shore receiving station performs error checking on each block as it is received, before storing it in memory. Retransmitted blocks identified as corrupted are noted in the table.

The process of requesting and retransmitting blocks continues until all the audio and video material has been transferred without error. The shore station confirms the successful completion of the transfer by repeatedly sending the Ack All packet for one minute to increase the probability of reception. This technique was used as it is the only message that does not initiate an action from the yacht system. This is illustrated in the time sequence diagram in Fig. 15.7.

15.7.3.4   YVS protocol — advantages

The ARQ protocol was designed specifically to handle the requirements of the YVS. It ensures that the channel is used for the transfer of data whenever possible (unlike 'stop-and-wait' protocols). The protocol overhead has been kept to a minimum (only 0.2% of total bandwidth) to ensure efficient use of the channel.

The protocol takes account of the 0.5 sec round-trip delay associated with satellite channels, as well as the long satellite link outages caused by the motion of the yacht.

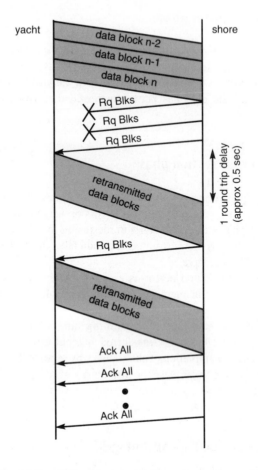

Fig. 15.7   YVS-error correction protocol — error-correction phase.

15.8   SYSTEM DESCRIPTION

Although similar components were used for both the yacht and shore-based systems, the yacht system required a highly integrated approach with special attention to environmental conditions, whereas the shore system was a standard office configuration.

The software for both systems was written in 'C' running under DOS V5.0. Figure 15.8 provides a diagram of the complete system.

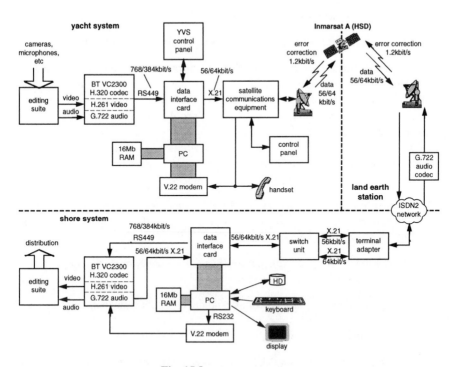

Fig. 15.8   System overview.

15.8.1   Yacht system

The yacht system was based around the H.320-compliant BT VC2300 video and audio codec. In addition to being the compression engine for the video and audio data, it also provided a useful enclosure for the other system components.

The codec was heavily modified to make the transition from the cosy confines of a videoconferencing suite to the harsh environment of a yacht. This included:

- soldering all electronic components into their sockets;

- hot gluing all connectors (enabled reworking if necessary);

- foam rubber strips on front panel fixed with stainless steel screws to provide additional support to internal boards;

- packets of desiccated silica gel strapped to top and bottom rails to absorb moisture;

- silicon-based water-resistant coating on all board surfaces;

- modification to card rails to prevent board dislocation under shock.

To assist further in the protection, a purpose-built shock and vibration damped cage was constructed to house the system. The cage also helped installation and waterproofing. Additional components (power unit and control panel) were housed in sealed enclosures, with waterproof cable glands to prevent internal contamination.

The store-and-forward aspects of the system were designed as a codec option card, with a custom-designed rear panel for connection to the other system components. This provided mounting and interconnection for the following:

- industrial PC104/ISA format 386SX PC with software contained in EPROM to give high reliability due to no moving parts, and 16-Mbyte RAM to store compressed video and audio data;

- V.22 modem for receiving retransmission requests from the shore system via the audio channel on the SES;

- data interface PC card custom-designed to enable the control panel and switch unit to determine the data flows between the PC, codec and communications equipment.

The data interface card was designed to accommodate both yacht and shore versions of the system. The RS449 interface to the codec enabled the video and audio data to be stored on the yacht and played back on shore. The X.21 port could be connected to the yacht's SES for data transmission or to an ISDN2 terminal adapter for on-shore reception. To enhance functionality further, the card was provided with interfaces for the yacht's control panel and the on-shore 56/64-kbit/s switch unit, designed to monitor and control system access.

Ideally, YVS control would have been via a PC keyboard and display situated in the yacht's navigation station. However, this area was already overcrowded with communication and navigation aids, including the control panel and telephone handset for the SES. The YVS control panel was therefore made small and simple, comprising:

- rotary switch to select 'store' or 'forward';

- toggle switches to 'start/stop', and select either '768 kbit/s' or '384 kbit/s' operation;

- tricolour LEDs to indicate status and progress;

- SES safety-system bypass key switch for emergency situations.

The YVS and SES required 110 V AC power supplied from a +24 V DC inverter. This placed demands on the yacht's power (185 W and 390 W respectively) and had implications on the quantity of diesel carried to power the generator. A power unit was designed to provide power switching between the YVS and the SES for the store-and-forward operations, while retaining power to the PC to preserve the video clip stored in memory.

To ensure the yachts carrying the system were not disadvantaged by the additional weight of the YVS and SES (30 kg and 65 kg respectively), lead ballast of equivalent weight was added to the unequipped yachts.

15.8.2   Shore system

The principal receiving station for the video clips was based at Reuters, London where it was running unmanned continuously throughout the duration of the race. A back-up system was kept at BT Laboratories in case of system failures, or congestion if more than one yacht wanted to transmit at the same time. The system components were:

- BT VC2300 video and audio codec in stand-alone office enclosure;

- 386DX desktop PC with 16-Mbyte RAM to store video and audio data temporarily on reception and playback, and 520 Mbyte hard disk for storing around 30 full-length video clips;

- V.22 modem for sending retransmission requests to the yacht system via the audio channel;

- data interface PC card (as used in the yacht system);

- ISDN2 terminal adapter providing two X.21 (56/64 kbit/s) ports;

- 56/64-kbit/s switch unit designed to monitor and control system access.

The codec formed the core of the shore station providing the means to decode the stored video and audio data streams back to analogue for subsequent recording and editing. Unlike the yacht system, the PC and modem were desktop items, with the data interface card located inside the PC.

V.22 modems were selected because 1.2 kbit/s was the maximum reliable data rate that could be achieved without needing a return channel for equalization purposes. Because the return audio channel only provided acknowledgements at the beginning of transmission and data requests at the end for error correction, 1.2 kbit/s was more than adequate.

As the video clips could arrive from land earth stations throughout the world, both 56-kbit/s and 64-kbit/s operation had to be supported. By defining the ISDN2 line to have 'separate channel numbering' each B channel was mapped to a unique number. Configuring the terminal adapter for B1 on port 0 (56 kbit/s) and B2 on port 1 (64 kbit/s) the SES operator could select which number to call depending on earth station and corresponding channel rate. When the call was made, the '56/64 kbit/s switch unit' sensed on which port of the terminal adapter the call had arrived and informed the system. The unit also asserted control lines on the terminal adapter to 'busy' the other port to prevent contention if another yacht attempted to call while engaged.

When not being used for the retrieval of the video clips, the codec was configured as a G.722 audio-only codec to convert the modem tones containing the error correction requests into a form suitable for the return channel on the ISDN2. The modem tones were converted back to analogue audio at the earth station prior to transmission.

15.8.3   PERFORMANCE

Before transmission, clips stored in the yacht system had to contain a 15 sec preamble to take account of the time taken for the shore system to synchronize to the video and audio data on retrieval. Ideally the preamble would be a static image such as colour bars or similar test pattern. The H.320 video and audio data stream does not have a defined start point, as its primary use is for videoconferencing in real time. In addition, the storing mechanism on the yacht depended on when the operator activated the 'start/stop' switch, which did not correspond to any particular point on the data's framing structure. When this data was applied to the shore system codec it had to synchronize at all levels before the video and audio data could be retrieved, causing loss of material.

On occasions the video clip received was not completely clear of errors due to premature termination of transmission. As the error-correcting system worked from the beginning of the clip to the end, this was not a major problem as the operators were advised to put key material first.

Problems were experienced with the SES failing due to the constant shock and vibration loads causing mechanical deterioration. Although comprehensive testing was performed beforehand the extent and severity of the operating environment stretched the system's construction to its limits. As the race progressed a comprehensive picture of failure modes was generated allowing corrective action at each port of call.

When the fleet entered the southern oceans, transmission of clips became difficult due to:

- extreme sea and weather conditions;

- low elevation angle to the satellite;

- flux-gate compass giving inaccurate bearings (magnetic deviation);

- signal degradation or loss through deck and hull when heeling;

- system operating on the edge of satellite coverage.

As no simple solution was available to overcome these problems it was basically up to the tenacity and patience of the operators to get the material off the yachts.

15.9   CONCLUSIONS

The Whitbread 1993 Round the World Race finished in June 1994 after nine months and 32 000 nautical miles of highly competitive sailing in extreme conditions. During this period a total of 238 minutes of prime broadcast material was successfully relayed by YVS on ten yachts from mid-ocean to the broadcasters world-wide, for inclusion into programmes such as ITV's 'Sail the World'. Although the systems were subjected to the heat of the equator and the cold of antarctic waters, the peace of the doldrums and the frenzy of the southern oceans, a constant flow of video material was achieved to bring the stories from mid-ocean alive. Problems were encountered by yachtsmen and engineers alike. However, the experience gained and the stretching of technology and networks will prove invaluable for the design of future store-and-forward systems using satellite communications.

15.10   REFERENCES

1.   International Telecommunication, Union Recommendation ITU-T H.320: 'Narrowband visual telephone systems and terminal equipment', (1993).

2.   International Telecommunication, Union Recommendation ITU-T G.722: '7 kHz Audio-Coding within 64 kbit/s', (1988).

INDEX